Strengthening European Climate Policy

"The tough still compelling need for the EU to achieve its climate neutrality ambitions can only benefit from the interdisciplinary and inclusive approach SSH can offer. I acknowledge the innovative insights for the design and implementation of effective and just climate policies that this book provides."
—Maria Kottari, *Principal Consultant Green Economy & Innovation, Technopolis Group*

Ester Galende Sánchez · Alevgul H. Sorman ·
Violeta Cabello · Sara Heidenreich ·
Christian A. Klöckner
Editors

Strengthening European Climate Policy

Governance Recommendations from Innovative
Interdisciplinary Collaborations

Editors
Ester Galende Sánchez
Basque Centre for Climate Change
Leioa, Vizcaya, Spain

Violeta Cabello
Basque Centre for Climate Change
Leioa, Vizcaya, Spain

Christian A. Klöckner
Department of Psychology
Norwegian University of Science
and Technology
Trondheim, Norway

Alevgul H. Sorman
Basque Centre for Climate Change
Leioa, Vizcaya, Spain

Sara Heidenreich
Department of Interdisciplinary Studies
Norwegian University of Science
and Technology
Trondheim, Norway

ISBN 978-3-031-72054-3   ISBN 978-3-031-72055-0 (eBook)
https://doi.org/10.1007/978-3-031-72055-0

© The Editor(s) (if applicable) and The Author(s) 2024. This book is an open access publication.

Open Access This book is licensed under the terms of the Creative Commons Attribution 4.0 International License (http://creativecommons.org/licenses/by/4.0/), which permits use, sharing, adaptation, distribution and reproduction in any medium or format, as long as you give appropriate credit to the original author(s) and the source, provide a link to the Creative Commons license and indicate if changes were made.
The images or other third party material in this book are included in the book's Creative Commons license, unless indicated otherwise in a credit line to the material. If material is not included in the book's Creative Commons license and your intended use is not permitted by statutory regulation or exceeds the permitted use, you will need to obtain permission directly from the copyright holder.
The use of general descriptive names, registered names, trademarks, service marks, etc. in this publication does not imply, even in the absence of a specific statement, that such names are exempt from the relevant protective laws and regulations and therefore free for general use.
The publisher, the authors and the editors are safe to assume that the advice and information in this book are believed to be true and accurate at the date of publication. Neither the publisher nor the authors or the editors give a warranty, expressed or implied, with respect to the material contained herein or for any errors or omissions that may have been made. The publisher remains neutral with regard to jurisdictional claims in published maps and institutional affiliations.

Cover illustration: © Melisa Hasan

This Palgrave Macmillan imprint is published by the registered company Springer Nature Switzerland AG
The registered company address is: Gewerbestrasse 11, 6330 Cham, Switzerland

If disposing of this product, please recycle the paper.

Foreword 1—Low-Carbon Approaches at the Crossroads: Why the European Green Deal Will Benefit from Interdisciplinary Insights

The European Union (EU) has outlined its ambitions to become the first climate-neutral continent. The achievement of this ambition is supported through the EU Green Deal which sets out a long-term roadmap to deliver on the long-term systemic changes required. The roadmap covers a range of activities across sectors, including climate, energy, and mobility. At the heart of the EU Green Deal is the commitment to put people first and leave no person (or region) behind.

The contribution of Social Sciences and Humanities (SSH) research to low-carbon transitions cannot be understated. I am firmly convinced that we will fail in delivering upon our climate neutrality ambitions without SSH. SSH contribute to low-carbon transitions in multiple ways, including the development of inclusive approaches, the establishment of effective communication, and the creation of appropriate governance structures.

SSH research supports the establishment of an inclusive approach to achieving the EU's climate neutrality ambitions. As mentioned previously, a key component of the EU's approach to achieving climate neutrality is that no one, and no region, is left behind. Inclusiveness is therefore important to ensure potential disparities and inequalities are addressed. SSH support the establishment of inclusive practices by providing insight into cultural factors, such as values, beliefs, and identities, and how these can support green policies and green transitions. Understanding the different cultures and experiences of individuals is a key component of ensuring no one is left behind.

SSH also support communication and inform the development of effective public engagement initiatives. The incorporation of SSH insights into the narratives of transition can help convey to the public how low-carbon transitions are beneficial for the planet, the health and well-being of individuals, and the economy. SSH also help demonstrate the necessity of behaviour changes to deliver on climate change. It is vitally important that policymakers get support on how to convey messages of urgency—but also the benefits on the lives of individual people—related to low-carbon policies. Not only does SSH research support effective communication related to low-carbon behaviours and practices, but it also provides insights on how changes will both be experienced and encouraged.

SSH research does not focus solely on behaviours and societal configurations; it also provides insights related to policy structures, institutions, and industry. These understandings can inform practices and help ensure that the actions undertaken are as effective as possible.

While SSH research and insights play a vital role in achieving the EU's low-carbon ambitions, they will have the most impact when integrated with other disciplinary perspectives, including those from Science, Technology, Engineering, and Mathematics (STEM). Most solutions for achieving low-carbon ambitions are situated at a crossroads: these solutions are not linked to a single sector or disciplinary background, rather they overlap between sectors and require the integration of different knowledge and perspectives. Bringing together different experts and experiences when developing approaches is essential to find innovative approaches to tackle climate change, undertake the energy transition, and establish sustainable mobility. Yet, in order to achieve all this, there is the need to continue breaking the silos in which research is undertaken and communicated. As such, interdisciplinarity between SSH and STEM needs to be promoted and supported.

Not only is there the need for interdisciplinarity between research disciplines, but there is also the need to have collaboration and communication between different actors. Achievement of low-carbon ambitions requires interactions between SSH, policymakers, and more technical and naturalistic disciplines. Policymaking needs to become more comprehensive and interdisciplinary in order to advance transitions, avoid duplications, and maximise impact through involving different people. The interactions between policy and research are critical as research and innovation activities need to be supported by the regulatory framework, with the regulatory framework also needing to be aware of the research and innovation activities undertaken to enable updates. Within policymaking, we need to continue to break those silos, adopt more interdisciplinary approaches, and make sure to bring along the societal dimension of the transition.

The collective intelligence across the three SSH CENTRE books—bringing together more than 150 researchers from more than 23 countries—is inspiring. The collaborations underpinning the chapters show how we should be working and are a starting point for breaking down silos. I really believe that collaborations between Social Scientists, Humanities researchers, and researchers from more technical disciplines, are key to advance low-carbon transitions. In order to achieve climate neutrality, there are many challenges to overcome, but the insights presented within these chapters and the expertise of the chapter authors can support the establishment of effective solutions, help break down barriers, and accelerate pathways to a sustainable and prosperous future.

Rosalinde van der Vlies
Brussels, Belgium

Rosalinde van der Vlies *is the Director of the Clean Planet Directorate in the European Commission's Directorate-General for Research and Innovation, and Deputy Mission Manager of the Climate-Neutral and Smart Cities Mission. Before her appointment as Director, Rosalinde van der Vlies was the Head of Coordination & Interinstitutional Relations Unit, and acting Head of Communication & Citizens Unit. Previously she held positions in Directorate-General Environment, Directorate-General Justice and Home Affairs, and in the private office of Janez Potočnik, the European Commissioner for the environment. Before joining the European Commission, she worked as a competition lawyer in an international law firm in Brussels and was a part-time teacher at the Catholic University in Brussels.*

Foreword 2—A Political Culture that Places Climate Change at the Centre of Social Priorities: Insights from an Interdisciplinary Approach

Some still argue that climate change is not a big deal and that those declaring it an existential threat to our current understanding of human existence are vastly exaggerating. I wish they were right. Unfortunately, the ever-increasing evidence insists on proving them wrong. For years, scientific reports have been piling up with stronger and starker warnings. Without an adequate response, the pernicious effects that climate change will bring will be impossible to reverse for generations. Impacts on our ecosystems, on our economy, but above all, impacts on our society as we know it. The IPCC, and with it the voice of science, is crystal clear: this decade is critical. Not only will it shape the immediate future but also the future of humanity for centuries to come.

Climate change is already upon us. This is why we must double efforts to halt it and adapt to the already underway effects. However, this will require undertaking profound changes and adopting innovative approaches to public policies. Only then will we be able to reduce risks and avoid additional costs and negative impacts. We need new ways of legislating, planning, and investing. In other words, we need to activate a new and transformative policy agenda to fight against climate change and, at the same time, bring progress and well-being to citizens. It is not a new beginning, much has been done during the last five years to mainstream climate goals into sectoral policies, paying special attention to the social aspects of these policies.

This is not an easy endeavour, not least because the needed climate policies will significantly disrupt current social behaviours and political inertia. This is particularly relevant in democratic societies, where no new policy will be feasible unless understood and socially accepted.

Contrary to what is often thought, there is wide social support for ambitious policies against climate change. Indeed, as per a recent global survey published in Nature (see Andre et al., 2024), climate change raises widespread concern across the globe, and climate policies are looked into with increasing interest.

In any case, society will quite rightly demand climate policies to be based on rigorous analysis. Although we can take advantage of the numerous contributions that academia has already made to this field, the formulation of climate change policies is still a young field, in many cases suffering from a sizeable analysis, reflection, and evaluation deficit.

This publication intends to contribute to these analyses and assessments by touching upon a wide range of crucial areas with a genuinely interdisciplinary approach. From environmental and social implications of climate strategies to marine spatial planning. From energy rehabilitation to community engagement in the just transition process. From adaptation to heatwaves to the integration of local knowledge into the EU's climate adaptation policymaking.

The analyses made in this publication highlight the need, and the difficulty, for climate policies being sensitive to inequalities. We must not forget that climate change affects the most disadvantaged more intensely. For this reason, it is essential to assess the social implications of every option and seek solutions to attain a fair distribution of the costs and benefits of the fight against climate change.

Moreover, this publication pays particular attention to adaptation policies, a relatively untapped field of climate action, which, however, poses particular challenges for policymakers. Among them:

- The strong dependence on local contexts, as formulas that have yielded positive results in a particular context may not necessarily be suitable for other socio-environmental contexts. Therefore, local and participatory approaches are particularly relevant in this area.
- The need for financial flows, as there is still a limited understanding of how to overcome the current lack of incentives to undertake these policies, even if the damages avoided have been proven to be larger than the cost that these policies involve.

- The difficulties in measuring their success in a scenario where the dangers they try to tackle evolve rapidly.
- Their enormous transversality, which requires coordinating the action of the multiple actors competent for areas touched by these policies (economic-wise, health-wise, environmental-wise…).

In any case, this enormous transversality is a feature that affects not only adaptation policies but also mitigation ones. It also reflects one of the great challenges for developing effective climate policies: the need to strengthen governance systems. Governance must be multisectoral, coherent across different territorial scales, and multi-actor.

I would like to end with a quote from the veteran climate scientist Mike Hulme who, in his splendid book *Why we disagree about climate change* (2009), states: *"As society has been increasingly confronted with the observable realities of climate change and heard of the dangers that scientists claim lie ahead, climate change has moved from being predominantly a physical phenomenon to be simultaneously a social phenomenon"*. When facing the climate crisis, we must create a political culture that places climate change at the centre of social priorities. I truly believe that, by providing proposals and raising crucial debates, this publication will be a valuable contribution to the creation of such a culture. I hope you enjoy it.

Valvanera Ulargui
Madrid, Spain

Valvanera Ulargui *is the Director General of the Spanish Climate Change Office at the Ministry for the Ecological Transition and Demographic Challenge. Previously, she was an advisor in the Infrastructure, Environment, Energy, and ITCs Division of the Ministry of Economy and Competitiveness, focusing on promoting the internationalisation of Spanish companies in climate change. From 2001 to 2013, she worked at the Spanish Climate Change Office, coordinating Spain's position in international climate negotiations and advising the General Director on climate change policies.*

References

Andre, P., Boneva, T., Chopra, F. et al. (2024). Globally representative evidence on the actual and perceived support for climate action. *Nature Climate Change*. 14, 253–259.

Hulme, M. (2009). *Why we disagree about climate change: Understanding controversy, inaction and opportunity*. Cambridge University Press.

Preface

The EU has committed to becoming the first climate-neutral continent by 2050, and this requires strengthening European climate research and policy through interdisciplinary collaborations resulting in governance recommendations that are both comprehensive and innovative.

How can science support climate policy in the EU, and help create more just and transformative futures? This book is built on the premise that the climate crisis is inherently a social crisis that requires SSH disciplines to collaborate with STEM fields along with other social actors and especially civil society towards true inter- and transdisciplinarity. There is growing recognition that social challenges and solutions need to be better embedded in technical recommendations, and vice versa. The importance of improving SSH-STEM interdisciplinary practice is therefore vital in the years to come.

As editors, we come from different interdisciplinary climate and environmental journeys spanning social and technical disciplines, including Environmental Sciences, Environmental Engineering, Psychology, Human Geography, Science and Technology Studies, and Political Sciences. We try to analyse relationships between political, economic, social, technical, and environmental issues contributing to an enhanced understanding of how individuals, particular actors, institutions, and power dynamics impact decision-making. Indeed, our own experiences have shown us there can be just as much diversity within SSH or within STEM as between one and the other.

This book is a core output from the Horizon Europe project SSH CENTRE: *Social Sciences and Humanities for Climate, Energy and Transport Research Excellence* that includes findings from novel interdisciplinary collaborations which were catalysed and funded by the project. Each chapter portrays a different complex problem of the social-ecological and climate crises and ways to tackle these challenges via interdisciplinary collaborations. The contributions in this book have been produced by collaborative teams that had, until now, never published together, and have developed policy recommendations by bridging disciplinary boundaries—these are not small tasks and we greatly appreciate the work that authors have undertaken. Our book on strengthening European climate policy is part of a three-volume collection, the other volumes focusing on energy and mobility policies. All three are available on open-access.

Fundamentally, this book offers new ways of reimagining our collective futures when dealing with a planet in polycrisis. The compilation of the chapters resembles "Un Mundo Donde Quepan Muchos Mundos" ("A World Where Many Worlds Fit") capturing diverse ways of understanding and envisioning **the world we live in and wish to transform towards**.

Leioa, Spain	Ester Galende Sánchez
Leioa, Spain	Alevgul H. Sorman
Leioa, Spain	Violeta Cabello
Trondheim, Norway	Sara Heidenreich
Trondheim, Norway	Christian A. Klöckner

Acknowledgments

This project is funded by the European Union's Horizon Europe research and innovation programme (under grant agreement no. 101069529) and by the UK Research and Innovation under the UK Government's Horizon Europe funding guarantee (grant no. 10038991).

We would like to acknowledge the Spanish Ministry of Science and Innovation through the Ramón y Cajal (RYC2021- 031626-I); the María de Maeztu programme for accreditation of excellence 2023–2027 (CEX2021- 001201-M); the Basque Government through Ikerbasque, the Basque Foundation for Science and the BERC 2022–2025 programme.

We additionally thank all the reviewers and their constructive feedback which played an invaluable role in enhancing the overall quality of the book.

We also sincerely thank Ami Crowther from Anglia Ruskin University (ARU) for coordinating the book series on Strengthening European Energy/Climate/Mobility Policy. Her support is greatly appreciated for making the entire editorial process seamless and efficient.

Competing Interests The editors declare that they have no competing interests that could have influenced the content, analysis, or interpretation presented in this book. Any affiliations with entities that could potentially be perceived as a conflict of interest have been disclosed transparently. The viewpoints and conclusions expressed herein are solely those of the editorial team and authors and do not necessarily reflect the official policy or position of any affiliated institutions or funding sources.

Contents

1. **Introduction** — 1
   Ester Galende Sánchez, Alevgul H. Sorman, Violeta Cabello, Sara Heidenreich, and Christian A. Klöckner

2. **Considering the Cross-Boundary Environmental and Social Implications of the EU's Carbon Dioxide Removal Strategy in Brazil** — 9
   Joana Portugal-Pereira, Aline Cristina Soterroni, Antonella Mazzone, and Jiesper Strandsbjerg Tristan Pedersen

3. **Weaving a Transformative Circular Textile Policy Through a Socio-Environmental Justice Lens** — 21
   Lis J. Suarez-Visbal, Martin Calisto Friant, Anna Härri, Veerle Vermeyen, Abe Hendriks, Blanca Corona Bellostas, and Jesus Rosales Carreon

4. **Adapting to Heatwaves: Reframing, Understanding, and Translating Strategies from India to the EU** — 35
   Laura Menatti, Anna-Katharina Brenner, Joyshree Chanam, Marina Knickel, Hari Sridhar, and Corey Bunce

5	Advancing Epistemic Justice with Local Knowledge: A Process Indicator for EU Climate Adaptation Policymaking Hernán Bobadilla, Giuseppe Di Capua, Chris Hesselbein, Silvia Peppoloni, and Federico Lampis	49
6	Linking Vulnerability to Heatwaves and Public Health: Indicators for EU Policies on Energy Renovation of Residential Buildings Ángela Lara-García, Carlos Rivera-Gómez, Claudia Núñez-Rivera, Carmen Galán-Marín, and Estrella Candelaria Cruz-Mazo	61
7	Reforming Carbon Accounting Mechanisms Around Justice-Based Principles to Promote Societal Sustainability Camilla Seeland, Piers Reilly, Ilaria Perissi, Diego Andreucci, Roger Samsó, and Jordi Solé	75
8	Leave No One Behind: Engaging Communities in the Just Transition Process Towards Climate Neutrality Ricardo García-Mira, Nachatter Singh Garha, Serafeim Michas, Franziska Mey, Samyajit Basu, and Diana Süsser	87
9	Developing Equitable Maritime Spatial Planning in the EU: Case Studies from Portugal and Norway Dina Abdel-Fattah, Misse Wester, Irene Martins, Sandra Ramos, and Stian K. Kleiven	99
10	Bringing in Ethics: A Multi-stakeholder Approach to Manage the Transition to Low-Carbon Construction Michal Plaček, Vladislav Valentinov, Roman Fojtík, František Ochrana, and Martina Peřinková	111
11	Integrating Multispecies Justice Approach for Climate Risk Management in Forest Areas of Mediterranean Europe Ethemcan Turhan, Cem İskender Aydın, Nurbahar Usta Baykal, and İsmail Bekar	125

| 12 | Conclusions | 137 |

Ester Galende Sánchez, Alevgul H. Sorman,
Violeta Cabello, Sara Heidenreich,
and Christian A. Klöckner

Afterword 1. Where Do We Go from Here? SSH Inquiries into Crucial HOW Questions in Climate Change Policies   145

Afterword 2. Cross-Disciplinary Thinking to Rise to the Challenges of Global Systemic Risks   149

Afterword 3. Justice as the Foundation of European Climate Policies: A Future that Serves All of Us   153

Afterword 4. Interdisciplinary Perspectives to Reimagine Systems for a Sustainable and Just Future   157

Index   161

Notes on Contributors

Dina Abdel-Fattah is an Associate Professor in the Department of Technology and Safety at The Arctic University of Norway and is the research group leader of UiT's Climate Change Adaptation research group. Abdel-Fattah has an established research background in climate change-induced hazards and their societal impacts, particularly in an Arctic context, leading several NordForsk, Erasmus+, and Norwegian Research Council projects.

Diego Andreucci is a Senior Research Fellow at the Universitat de Barcelona. He received his PhD from the Institute for Environmental Science and Technology at the Autonomous University of Barcelona and holds a Bachelor's degree in Philosophy and Anthropology from the University of Rome "La Sapienza", and a Master's in Geography from the National University of Ireland, Galway. His research interests include Political Economy and Ecology; Environmental Governance and conflicts; and Indigenous and Environmental Movements.

Cem İskender Aydın is an Assistant Professor at the Institute of Environmental Sciences in Boğaziçi University, Turkey. His research covers the topics of energy justice, energy and climate policy/politics, environmental justice, and environmental valuation. He was also a fellow for IPBES Values Assessment published in 2022.

Samyajit Basu is a Senior Researcher in the Mobilise research group at Vrije Universiteit Brussel (VUB). His primary research interests are sustainable transportation, transport policy, road safety, and human factors and just transition in transportation. He has worked in and coordinated multiple Horizon 2020 and Horizon Europe projects on these themes. His past research engagements also include environmental impact assessment projects in India and driving simulator studies in Italy.

İsmail Bekar is currently a Postdoctoral Researcher at the Technical University of Munich with an interest in various aspects of wildfire research. He is an ecologist with a PhD from ETH Zurich focusing on understanding the intricate dynamics within fire regimes.

Hernán Bobadilla is an Assistant Professor at the Department of Mathematics at the Politecnico di Milano. He is a philosopher of science and geologist who studies epistemological issues in the sciences, e.g., surrogative reasoning with scientific models, scientific explanations, and understandings. He holds a PhD in Philosophy from the University of Vienna.

Anna-Katharina Brenner (BA, MSc, soon PhD) is a Research Associate at the Institute of Social Ecology at the University of Natural Resources and Life Sciences. She investigates transformative actions that reuse built environments for sustainability. Her focus is on the systemic relationships between actors, institutions, and materials that shape interventions. She develops and applies reflexive, integrated mixed-method approaches, combining natural science modelling (GIS), social science, and art-based research.

Corey Bunce is a Postdoctoral Fellow at the Konrad Lorenz Institute for Evolution and Cognition Research, where he studies scientific discourse using literary theory and semiotics.

He is a developmental biologist and philosopher of science. He received a PhD in Developmental Biology from Duke University where he researched organogenesis and sex determination.

Martin Calisto Friant is a Postdoctoral Researcher at the Autonomous University of Barcelona. He lectures on circular economy at the University of Amsterdam and EADA Business School. He has an interdisciplinary academic background in International Development, Political Ecology, Environmental Governance, and Urban Studies from Utrecht University (PhD), the University of Melbourne (MA), University College London (MSc), and McGill University (BA). He has eight years of experience as a sustainability practitioner and researcher working on projects in Europe, Oceania, Africa, and North and South America.

Estrella Candelaria Cruz-Mazo is an Assistant Professor at the Faculty of Geography and History (University of Seville). She is also a visiting researcher at the Autonomous University of Madrid, the Paris Urban Planning Institute, and the OTB Research of the Built Environment at the Delft University of Technology. Her research interests are focused on housing and rehabilitation policies and residential vulnerability.

Violeta Cabello is a Ramón y Cajal Researcher at the Basque Centre for Climate Change (BC3). Her research interests include knowledge co-production, social-ecological systems, water governance, and environmental conflicts. She feels moved by questions around the "how(s)" of living together on a damaged planet.

Joyshree Chanam holds a PhD in ecology from the Indian Institute of Science, Bangalore. Her research interests include cost and benefits of mutualisms, the impacts of global warming on plant-pollinator interactions, and more recently the human perceptions and adaptations to climate change and heatwaves using an oral history approach.

Blanca Corona Bellostas is an Assistant Professor at the Copernicus Institute of Sustainable Development at Utrecht University. She is an Industrial Design graduate from the University of Zaragoza and obtained her Master's and Doctorate degrees in Environmental Engineering at Universidad Politécnica de Madrid (UPM). Her main research expertise is related to the sustainability and circularity assessments of products and services. She is an experienced researcher in the life-cycle sustainability assessment of innovative technologies and products, covering environmental, social, and economic impacts.

Giuseppe Di Capua is a Geologist at the Italian Institute of Geophysics and Volcanology, specialising in seismic hazard, geoethics, and social geosciences. He is a Member of the International Union of Geological Sciences Geoethics Commission, an Executive Committee Member of the International Council for Philosophy and Human Sciences, and the co-founder of the International Association for Promoting Geoethics.

Roman Fojtík is an Associate Professor of Civil Engineering and Biomaterials at the Technical University of Ostrava. He is very strongly connected to the practice as an authorised designer and expert witness, specialising in transport construction. He is the author of the longest wooden-concrete bridge in Central Europe. He and his team also won the Green Lights startup competition with an innovative energy storage system.

Carmen Galán-Marín is a Full Professor of Construction Technology and Sustainable Architecture at the Faculty of Architecture of the University of Seville, where she leads a multidisciplinary group involved in urban climate adaptation, linking different urban intervention strategies to the outdoor thermal comfort perception by citizens.

Ester Galende Sánchez is a Researcher at the Basque Centre for Climate Change (BC3). She has an interdisciplinary background in Environmental Engineering, Communication studies, and Sustainability, and holds a PhD in Political Science. Her professional and research interests lie in climate policy and governance at the European and international levels, with a focus on citizen participation and justice.

Ricardo García-Mira is a Professor of Social Psychology at the University of A Coruna, where he led the People-Environment Research Group from 1995 to 2022. He was an honorary Professor at the Institute for Policy Research of the University of Bath (2016–2024) and President of the International Association People-Environment Studies (2014–2018). He is a Fellow of the International Association of Applied Psychology.

Anna Harri is finishing her PhD at LUT University in Finland. Her dissertation focuses on the just transition to a circular textile economy, in the context of the Global South. She has also recently started as a Postdoc at the Aalto University Business School, where she researches Sustainable and Regenerative Forest Management.

Sara Heidenreich is a Senior Researcher at the Norwegian University of Science and Technology, where she leads the "Centre for Climate, Energy and the Environment". Her research interests include Sustainability Transitions and Climate Adaptation, with a particular focus on Participation and Justice, using perspectives from Science and Technology Studies (STS).

Chris Hesselbein is an Assistant Professor at the Politecnico di Milano. He is an ethnographer who studies how knowledge and technology are co-constructed with conceptions of social order, particularly on the level of embodiment, materiality, and aesthetics. He holds a PhD in Science and Technology Studies from Cornell University.

Abe Hendriks is an Assistant Professor at Utrecht University, where he focuses on the Politics and Power in imagined futures. In his PhD research, Abe studied how the circular economy is imagined as a desirable future in different regions in Sweden in the Netherlands. His main research interests relate to the interaction between Science, Innovation, and Society, especially in the context of Sustainability Transitions. He is concerned with how desirable futures are collectively imagined and how these imaginaries of desirable futures are inherently connected to our understanding of present-day sustainability-related problems.

Stian K. Kleiven is a PhD Researcher specialising in Aquaculture and Marine Ecology at The Arctic University of Norway. His research focuses on Climate Change Impacts and Adaptation within marine systems. Stian is also an active member of the research group Climate Change Adaptation and Changing Arctic Research School.

Christian A. Klöckner is a Professor in Social and Environmental Psychology at the Norwegian University of Science and Technology, where he leads the research group for "Citizens, Environment, and Safety". His main research interests are drivers and barriers of pro-environmental behaviour in its social and physical context and innovative environmental communication.

Marina Knickel is a Social Scientist working on transdisciplinary approaches in agri-food and land-use research. Her current work lies at the intersection of transdisciplinary Sustainability research, Social Sciences, and Humanities. She is interested in strengthening science-society collaborations by focusing on mutual learning for actors' capacity building, knowledge integration, and epistemic justice.

Federico Lampis is a Policy Officer at the European Commission's Directorate-General for Research and Innovation. Before joining the Commission, he worked at the OECD's Public Affairs and Communications Directorate and as a public affairs specialist in a Brussels-based consultancy firm. He has a Master's in European Affairs from Sciences Po Paris.

Ángela Lara-García is a Lecturer at the Department of Human Geography of the University of Seville. She is an Architect and holds a PhD in Geography. Her research focuses on the interaction between urban land and its socio-ecological context, with a specific emphasis on vulnerability assessment and climate change adaptation strategies, exploring nature-based solutions to mitigate environmental risks.

Irene Martins leads CIIMAR's Marine Ecosystem Modelling team at the University of Porto. Her work involves crafting ecological models to comprehend and predict marine ecosystem dynamics amid stressors such as climate change and pollutants. She advocates for multidisciplinary approaches, merging various numerical tools and actors to promote sustainable marine ecosystem management.

Antonella Mazzone a Research Associate at the Department of Anthropology and Archaeology (University of Bristol) and at the Centre for the Environment (University of Oxford). She is a Social scientist specialising in decolonial-feminist-intersectional perspectives on extreme heat and cooling poverty in marginalised communities in Latin America and the Caribbean region.

Laura Menatti is a Senior Postdoctoral Fellow at the Konrad Lorenz Institute for Evolution and Cognition Research. She is a philosopher of science dedicated to exploring the theoretical underpinnings and practical implications of the relationship between health and the environment. She is investigating the health impacts of environmental factors within the fields of biomedical and environmental sciences. Currently, she is actively engaged in studying climate change adaptation, focusing on conceptualisation, applications, and strategies for innovation.

Franziska Mey co-leads the research group "Democratic Governance and Agency" at the Research Institute for Sustainability (RIFS) Helmholtz Centre Potsdam. With master's degrees in Political Science and Regional Development, she completed her doctoral studies in Community Energy. Her research now centres on Participation, Public perceptions, and Justice within the energy transition and regional transformation processes.

Serafeim Michas is a Research Associate at the Technoeconomics of Energy Systems Laboratory at the University of Piraeus. He has a PhD in the field of Technoeconomics of Energy Systems with a focus on energy modelling. He has eight years of research experience, participating in several European and national funded projects.

Claudia Núñez-Rivera is a PhD Researcher in Geography at the University of Seville, specialised in Vulnerability and Natural hazards, with the aim of developing a methodology based on a proposal of dynamic indicators for the measurement of vulnerability at a finer scale.

František Ochrana is a Senior Researcher and Full Professor at Charles University. His research fields are Public Administration and the Methodology of Science. He has published a number of scientific articles in high-ranking journals (e.g., Governance, Environmental Sciences Europe, Waste Management, and Public Money & Management, among others), several dozen books, and has been involved in numerous scientific projects. He regularly serves as an expert for several central government bodies and is a member of the International Institute of Public Finance.

Silvia Peppoloni is a PhD Researcher at the Italian Institute of Geophysics and Volcanology, specialising in Georisks, Geoethics, and Social Geosciences, holding positions as Chair of International Union of Geological Sciences Geoethics Commission, Chairholder on Geoethics of International Council for Philosophy and Human Sciences, and Advisory Board member for Climate Intervention Research of American Geophysical Union.

Martina Peřinková is a Professor of Architecture at the Technical University of Ostrava. She is a practising architect and in her private practice she designs civic amenities and buildings for housing. A significant part of her work involves working with the public space of towns and villages, and interior design is also a stable part of her architectural studio.

Joana Portugal-Pereira is a Professor at the IST/ULisboa and Coppe/UFRJ (Brazil). She holds a PhD in Urban Engineering from The University of Tokyo (2011). She contributed as an author to the IPCC Sixth Assessment Cycle and the UN Environment Programme's Emissions Gap Report. She is specialised in energy-land-based innovations, and develops tools for assessing socio-economic-environmental co-benefits in low-carbon trajectories.

Piers Reilly is a PhD Researcher at Anglia Ruskin University, having previously worked in government and industry, advising on Sustainability, Security, and International Law. His multidisciplinary research examines the Governance of the Net-Zero Transition. This involves the creation of an alternative governance and policy framework, designed to deliver a rapid transition to net-zero carbon throughout a region. His PhD blends participatory action research strategies, transition management methods, and techniques drawn from alternative dispute resolution.

Carlos Rivera-Gómez is a Chair Professor at the School of Architecture of the University of Seville. His main lines of research are focused on Sustainable Architecture, Urban microclimates, Indoor and Outdoor Thermal Comfort, Sustainable Materials and Products, and Energy-Saving Assessment.

Jesus Rosales Carreon is an Assistant Professor at the Copernicus Institute of Sustainable Development at Utrecht University. He is a Chemical Engineer from the National Autonomous University of Mexico. He has an MSc in Energy and Environmental Sciences and a PhD from Groningen University in the field of Knowledge Management in Sustainable Agriculture. He is interested in understanding the environmental and social implications associated with the implementation of the concept of circular economy and living labs. Current projects are embedded in the fashion, construction, and education sectors.

Ilaria Perissi is a Research Fellow at the Geosciences and Earth Resources Institute of National Research Council (Italy), where she investigates the potential of wetlands as carbon sinks. She was previously a Marie Curie Fellow at Anglia Ruskin University with the PLEDGES project, which aimed to develop a new climate policy modelling tool managing carbon budget targets for EU-27 countries. She is also an Associate Member of The Club of Rome, and an active promoter of sustainable cycling.

Michal Plaček is an Associate Professor of Public and Social Policy at Charles University and a Research Associate at the Global Institute of Sustainability and Innovation at the Arizona State University. His research is focused on green public procurement and cross-sectoral cooperation. He regularly publishes in recognised journals like Public Management Review, Public Money and Management, etc. His work is often cited by international organisations like OECD, IMF, or the European Commission.

Sandra Ramos leads the CIIMAR's Fish Ecology and Sustainability team at the University of Porto. Her primary research area is Marine Ecology, specifically concerning how human activities, including marine litter and microplastics, impact ecosystems and their ability to provide services and societal benefits.

Camilla Seeland is a PhD Researcher at the Global Sustainability Institute at Anglia Ruskin University. Her research explores the role of local government in driving the net-zero transition. She is specifically looking at the relational power dynamics between local policy officers and local micro and small-scale business owners and procedural justice. Before commencing on this PhD journey, she acquired a Master of Arts in Geography and International Relations and Master of Science in Environmental Partnership Management from the University of Aberdeen.

Roger Samsó is an Associate Lecturer at the Universitat de Barcelona. They obtained their PhD in Environmental Engineering from the Universitat Politecnica de Catalunya. Their research interests encompass diverse areas, including integrated assessment models and broader environmental sustainability issues. With expertise in modelling and scientific programming, they contribute significantly to interdisciplinary research and innovative solutions.

Nachatter Singh Garha is a Postdoctoral Researcher at the University of A Coruña. He holds a PhD in Demography from the Centre for Demographic Studies and the Autonomous University of Barcelona (UAB). He is working on an EU project TRIGGER studying the impact of climate change on health.

Jiesper Strandsbjerg Tristan Pedersen is a Climate Researcher and Consultant with a PhD in Emission Scenarios and Science-Policy Communication from Utrecht University, and a Master's in Anthropology from Copenhagen University. He is an Invited Lecturer at the Universities of Lisbon and Copenhagen. He is a member of eC3c-CHANGE research unit, developing Portuguese policy scenarios for the RNA2100 Project.

Jordi Solé is an Associate Professor at the University of Barcelona and Associate Researcher at CREAF. He obtained a Degree in Physics from the University of Barcelona, and a PhD in Applied Physics from the Universitat Politècnica de Catalunya. His research experience is focused on the study of ocean interactions between Physics and Biology and on Energy Modelling.

Alevgul H. Sorman is an Ikerbasque Research Associate at the Basque Centre for Climate Change (BC3). Her research scrutinises energy transitions with insights from human geography, political ecology, and ecological and biophysical economics focusing on issues of labour, space, gender, justice, and participation.

Aline Cristina Soterroni is a Research Fellow at the University of Oxford, working on scenarios modelling for environmental and climate policy evaluation in Brazil. She explores interactions between Net-Zero Policies and Nature-based Solutions. Soterroni advises Brazil's Ministry of Science, Technology, and Innovation while collaborating on a modelling platform for climate and environmental policy evaluation.

Hari Sridhar is a Senior Fellow at the Konrad Lorenz Institute for Evolution and Cognition Research and oversees the Oral History Programme of the Archives at NCBS (National Centre for Biological Sciences). Hari's current research examines the intersection of Ecology and Conservation Practice in India, through oral histories of conservation scientists.

Lis J. Suarez-Visbal is an Ashoka fellow and a PhD Researcher at Utrecht University. She has more than fifteen years of professional experience in supply chain sustainability in the textile and apparel sector. She has a BS in Finance and International Development from the Externado University of Colombia and an MSc in Sustainable Business and Innovation from Utrecht University. Her field of expertise is circular economy, system change, and social impacts in highly feminised value chains.

Diana Süsser is a Senior Expert at the Institute for European Energy and Climate Policy (IEECP). She serves as the coordinator of the Life project JUSt Transitions and EMpowerment against energy poverty (JUSTEM). She is an interdisciplinary researcher with a PhD in Geography. Her research interests lie in the field of policies to facilitate just transitions away from coal and in the meaningful engagement of the public in energy transitions.

Ethemcan Turhan is an Assistant Professor of Environmental Planning at the University of Groningen. His main research interests are situated at the intersection of Climate Justice and Energy Democracy with empirical attention to Environmental Movements. He edited *"Urban Movements and Climate Change: Loss, Damage and Radical Adaptation"* (Amsterdam University Press, 2023).

Nurbahar Usta Baykal is a PhD Researcher at the Hacettepe University, Division of Ecology. Her research is shaped around the effects of climate change in forest ecosystems; both in terms of community dynamics and species specific responses. She is currently a visiting researcher at Centro de Investigaciones sobre Desertificación, Spain.

Vladislav Valentinov is a Senior Researcher at the Leibniz Institute of Agricultural Development in Transition Economies and a Professor at the Department of Law and Economics at Martin Luther University. He is an expert on Institutional Economics and Systems Theory approaches to the third sector. A large part of his third sector research activities was supported by the Marie Curie International Fellowship of the European Commission and the Schumpeter Fellowship of the Volkswagen Foundation.

Veerle Vermeyen is a joint PhD Researcher between the KU Leuven and the University of Utrecht. Her research is focused on assessing the potential environmental gains the transition towards a circular economy for consumer textiles could entail. Her research methods are mainly from the field of industrial ecology, such as stock-flow modelling and life-cycle assessment, but also venture into consumer behaviour.

Misse Wester is a Professor of Risk Management and Societal Safety at Lund university, Sweden. She has broad experience with research on risk perception, risk communication, and crisis management related to climate change. With a PhD in psychology, she has worked extensively with exploring the links between cognition and behaviour.

List of Figures

Fig. 2.1	Analytical framework of this study (*Source* Own elaboration)	12
Fig. 3.1	Interdisciplinary analytical framework combining five interrelated and interconnected dimensions of socio-environmental justice (*Source* Modified from Härri and Levänen [2024, 14]	26
Fig. 4.1	Interdisciplinary adaptation framework as a magnifying lens which we used to understand adaptation to climate change in India. Following examination of multiple "Adaptation strategies" for extreme temperatures, a series of "Learning possibilities" for the EU emerged which highlight crucial "Approaches to global interrelations". Graphic by Corey Bunce	43
Fig. 5.1	Epistemic justice as composed of distributive, participatory, and recognitional justice (cf. Mathiesen, 2015)	54

List of Tables

Table 6.1	Indicators of residential vulnerability to heat (VH)	67
Table 7.1	Workshop participants' collective decisions and suggested social datasets	81
Table 10.1	Overview of key stakeholders and ethical trade-offs	116

CHAPTER 1

Introduction

Ester Galende Sánchez⊙*, Alevgul H. Sorman*⊙*,
Violeta Cabello*⊙*, Sara Heidenreich*⊙*,
and Christian A. Klöckner*⊙

Abstract The climate crisis represents a unique, collective action problem requiring the collaboration of diverse knowledge systems for enhancing climate policy. To achieve this, the book proposes the following: (1)

E. Galende Sánchez (✉) · A. H. Sorman · V. Cabello
Basque Centre for Climate Change (BC3), Leioa, Vizcaya, Spain
e-mail: ester.galende@bc3research.org

A. H. Sorman
e-mail: alevgul.sorman@bc3research.org

V. Cabello
e-mail: violeta.cabello@bc3research.org

S. Heidenreich
Department of Interdisciplinary Studies of Culture, Norwegian University of Science and Technology, Trondheim, Norway
e-mail: sara.heidenreich@ntnu.no

C. A. Klöckner
Department of Psychology, Norwegian University of Science and Technology, Trondheim, Norway
e-mail: christian.klockner@ntnu.no

© The Author(s) 2024
E. Galende Sánchez et al. (eds.), *Strengthening European Climate Policy*, https://doi.org/10.1007/978-3-031-72055-0_1

Supporting SSH-STEM collaborations to foster innovative research and holistic policies; (2) Integrating local knowledge and epistemic justice to ensure inclusive and effective policymaking; (3) Ensuring just and inclusive transitions that advance distributive, procedural, and restorative justice, prioritising community engagement and addressing social vulnerabilities; (4) Learning from the Global South for climate adaptation strategies; and (5) Strengthening climate adaptation through ethical governance which is value-based and locally grounded. Achieving these goals demands profound societal transformations as presented in this collaborative work of Strengthening European climate policy: Governance recommendations from innovative interdisciplinary collaborations.

Keywords Climate crisis · EU Green Deal · Just transitions · Interdisciplinary research · SSH-STEM collaboration

The climate crisis is a unique **collective action problem** in terms of scale, complexity, and severity (Bäckstrand & Lövbrand, 2015). These problems require collective approaches, diverse knowledge systems, and therefore novel collaborations to confront a world in polycrisis.

Faced with the climate crisis, among others, the EU has committed to becoming the first climate-neutral continent by 2050 with multiple strategies, initiatives, and directives being developed to support this shift. The EU Green Deal provides a roadmap for achieving the EU's climate neutrality ambition, outlining priority areas of action such as enforcing a reduction of 55% of their greenhouse gas emissions (GHG) by 2030 (compared to 1990 levels), supporting just and inclusive transitions that leave no one behind, and protecting and restoring ecosystems and biodiversity with more efficient resource use. The EU has also launched an ambitious Adaptation Strategy to become climate resilient by 2050 through smarter, faster, and systemic policy actions.

The achievement of these ambitions requires a complete transformation in the ways in which societies work, operate, mitigate, and adapt to the climate crisis. Social transformation is a complex process operating at different levels and spatiotemporal scales that require active and synergistic bridges (O'Brien et al., 2023). These bridges demand new forms of knowledge capable of breaking traditional silos once and forever to offer a comprehensive understanding and guidance on the challenges ahead

(Fazey et al., 2020; Mauser et al., 2013). As a response, the number of inter- and transdisciplinary collaborative research projects has multiplied, looking for innovative ways to engage with diverse knowledge domains and conduct meaningful and people-centric research for transformative change (Augenstein et al., 2024).

In this edited volume, we present 10 interdisciplinary contributions forming unique SSH-STEM collaborations to strengthen European climate policy and governance. They include a range of different methodological approaches, from more quantitative ones (e.g., complex system models, multivariate and multi-criteria analysis) to qualitative and participatory ones (e.g., collaborative workshops, interdisciplinary seminars), as well as novel combinations of both. There are deep philosophical inquiries on whether everyone understands ethics and justice in the same way, beyond anthropocentric framings. Several chapters question whether the persistent economic growth paradigm needs to be fundamentally challenged and how the EU can establish relations of mutual learning with the Global South. There are central critiques on power dynamics and calls for decentralisation across stakeholders, Global North-Global South actors, and beyond—to include more-than-human worlds as well.

The book is organised around three main parts: #1 Learning from the Global South for EU Climate Policy; #2 Measuring and Advancing Justice in EU Climate Policy; and #3 New Frontiers: Exploring Themes in Climate Policy.

In the first part, we present three chapters focused on learning from the Global South where practical strategies for climate adaptation have been in place for centuries. This part emphasises that a vision beyond Eurocentrism is needed considering that the EU Green Deal might have disproportionate effects and unintended consequences beyond its borders.

Portugal-Pereira et al. (Chapter 2) focus on the overreliance of the EU on offshore Carbon Dioxide Removal strategies and its impacts in Brazil mirrored through deforestation, biodiversity loss, land-use conflicts, and the erosion of indigenous peoples' rights and traditional knowledge. The chapter deploys a mixed-method analysis with the GLOBIOM-Brazil model and looks at environmental and social threats from Eucalyptus plantations in a monoculture regime risking additional land required in Brazil that is needed to meet the EU's net-zero targets. Suarez-Visbal et al. (Chapter 3) focus on the impact of current overproduction and overconsumption of the textiles sector emphasising that the majority of textiles in the EU originate from the Global South. After an in-depth review of 11 selected EU policies and 25 actions using an interdisciplinary framework combining SSH and STEM methodologies, the authors conclude that current circular economy policies fall short of justice standards and lack adequate recognition of actors from the Global South. The authors call for a transformative Circular Economy policy encompassing dimensions of recognition, distributive, procedural, and restorative justice. With this detailed policy analysis, they not only deliver an important input to a redefinition of circular economy concepts but also a methodology to check policy in other domains against global justice challenges. Menatti et al. (Chapter 4) then take us to the impacts of extreme heat in India and draw attention to reframing adaptation to encompass situated, relational, and longer-term processes of improving social-ecological adjustments involving a dialogue across local and global scales. The authors propose an interdisciplinary framework for the translation of long-term adaptation strategies in the South within the EU Adaptation Strategy, such as reviving traditional climate-sensitive architecture, embracing sacred trees as social spaces and Nature-Based Solutions, and mechanisms for empowering women in climate-friendly solutions.

All three chapters make calls for **accountability, responsibility, and empowerment**: either through responsible carbon offsetting, and alternative practices focusing on regenerative agriculture and restorative forestry (Chapter 2), elevating the voices, interests, and visions of the most marginalised people working in textile production in the Global South (Chapter 3) or promoting a culture of mutual learning for adaptation between the EU and global South countries in (Chapter 4). This calls for a rethinking of climate action respecting socio-spatial vulnerabilities and embracing a relation of mutual learning and respect between different geographies.

In the second part, we present four chapters that advance the idea of **justice** in climate policy, imperative not only for ensuring that no one is left behind but also advance effective and inclusive policies of the EU Green Deal.

Bobadilla et al. (Chapter 5) call for the integration of local knowledge as well as the advancement of epistemic justice in the EU Adaptation Strategy. The authors propose a process indicator to assist the policy-making process in two phases: the ex-ante (the problem framing) and the ex-post (the appraisal of the policy design). This indicator is set to advance distributive, participatory, and recognitional epistemic justice in adaptation but also in other EU policies, aiming to address existing gaps in the EU Better Regulation framework. Lara-García et al. (Chapter 6) draw attention to recognising heatwave vulnerability both as a climate and a prime public health problem, calling for an energetic rehabilitation of buildings to focus primarily on contexts of vulnerability, a foremost challenge in the Mediterranean. The authors propose a redefinition of heat vulnerability attending to social, environmental, and building factors such as energy poverty using a multi-criteria framework for its assessment. They apply this framework to the focus on the case of Seville, where 31.8% of heat-related deaths are caused by climate change, through a mixed-method approach: a multidisciplinary, expert workshop supported with a multivariate analysis. This framework can be used to prioritise European fund distributions within the Renovation Wave Strategy and the recently adopted Energy Performance of Buildings Directive. Seeland et al. (Chapter 7) focus on building justice-based carbon accounting mechanisms, highlighting the need to move away from purely emissions-based decision-making to one that also considers societal benefits. Combining Input–Output analysis with a collaborative workshop, the authors build a sustainability index that portrays trade-offs between the environmental impacts and societal well-being (e.g., employment, health, etc.). They introduce a concrete four-step plan for policymakers to achieve proactive, equitable, and targeted interventions towards carbon reduction. García-Mira et al. (Chapter 8) highlight the vitality of community engagement in transitioning coal and carbon-intensive regions, although reveal that local communities do not feel sufficiently engaged in the ongoing Just Transition Mechanism. The authors draw on the insights from four European research projects to reflect on the importance of tacit knowledge and socio-cultural configurations and highlight the need to support community energy projects and the development of visions, plans, and narratives at the local level for true transformation potential to unravel. Furthermore, they also show how modelling tools from STEM disciplines pose a golden opportunity to start a societal negotiation of transition pathways.

In this part, all four chapters draw on novel theoretical means and practical measures to further implement justice mechanisms in EU Climate Policy, all with a strong focus on social vulnerability. Chapters 5–7 propose alternative metrics to capture a more complex and nuanced vision of development, one that centres on human well-being and justice elements at its core, while Chapter 8 emphasises the importance, yet the recurrent lack, of community engagement in low-carbon transitions, and the need for novel tools to make this transition process really inclusive.

In the third part, we present three chapters that cover different areas and themes that are new frontiers to consider in **climate adaptation** policy and governance.

Abdel-Fattah et al. (Chapter 9) focus on land-and-sea interactions and draw attention to Maritime Spatial Planning as a decision-making tool and conflict resolution resource when confronted with issues such as deep-sea mining. By combining the insights from Social Sciences, Marine Ecology, and Climatology, the authors present two key case studies from Portugal and Norway to highlight how Maritime Spatial Planning could be a tool for the just and equitable distribution and use of maritime areas. They emphasise the importance of inclusive stakeholder participation processes that take into account the principles of distributional, recognition, and procedural fairness. Plaček et al. (Chapter 10) draw attention to ethical trade-offs in the construction sector (e.g., profit motives, conflicts of interests, accounting and integrity, corruption, privacy, and advertising) and draw insights from a participatory workshop managing the interests of different stakeholders within the supply chain, considering the risk of vulnerable stakeholders such as material producers and looking into ethical precautions in public procurement. The authors propose inclusivity in design and decision-making across the supply chain and accountability through transparency, highlighting the important role of public sector leadership in ethical transitions. Lastly, in the face of EU-wide wildfire control, Turhan et al. (Chapter 11) call for locally grounded, value-based response mechanisms (e.g., IPBES's Nature's Contribution to People framework) for climate change adaptation in Euro-Mediterranean forests. The authors bring together experts and communities to create synergies across local ecological knowledge and traditional and scientific practices for deterrence-focused firefighting strategies.

These three chapters reflect on natural (marine, forest) or artificial (building) spaces as heavily impacted socio-ecological systems suffering the consequences of the climate crises. Each one calls for recognising the multiplicity of actors, and knowledge claims brought forth by each for strengthening climate policy and governance.

This book can be read as: (i) stand-alone chapters, (ii) through common and cross-cutting themes that have been indexed, or (iii) as an evolving storyline, from looking at unbalanced global dynamics and unsustainable patterns (Part I) to diagnosing metrics and centring around justice elements (Part II) to advancing climate knowledge and policy into new frontiers (Part III).

As the editors of this collection, our hope is that novel SSH-STEM collaborations and interdisciplinary work bring forth a more holistic understanding and collaborative praxis of the many challenges and collective action problems that await.

References

Augenstein, K., Lam, D. P., Horcea-Milcu, A. I., Bernert, P., Charli-Joseph, L., Cockburn, J., ... & Sellberg, M. M. (2024). Five priorities to advance transformative transdisciplinary research. *Current Opinion in Environmental Sustainability, 68*, 101438.

Bäckstrand, K., & Lövbrand, E. (Eds.). (2015). *Research handbook on climate governance*. Edward Elgar Publishing.

Fazey, I., Schäpke, N., Caniglia, G., Hodgson, A., Kendrick, I., Lyon, C., ... & Saha, P. (2020). Transforming knowledge systems for life on Earth: Visions of future systems and how to get there. *Energy Research & Social Science, 70*, 101724.

Mauser, W., Klepper, G., Rice, M., Schmalzbauer, B. S., Hackmann, H., Leemans, R., & Moore, H. (2013). Transdisciplinary global change research: The co-creation of knowledge for sustainability. *Current Opinion in Environmental Sustainability, 5*(3–4), 420–431.

O'Brien, K., Carmona, R., Gram-Hanssen, I., Hochachka, G., Sygna, L., & Rosenberg, M. (2023). Fractal approaches to scaling transformations to sustainability. *Ambio, 52*(9), 1448–1461.

Open Access This chapter is licensed under the terms of the Creative Commons Attribution 4.0 International License (http://creativecommons.org/licenses/by/4.0/), which permits use, sharing, adaptation, distribution and reproduction in any medium or format, as long as you give appropriate credit to the original author(s) and the source, provide a link to the Creative Commons license and indicate if changes were made.

The images or other third party material in this chapter are included in the chapter's Creative Commons license, unless indicated otherwise in a credit line to the material. If material is not included in the chapter's Creative Commons license and your intended use is not permitted by statutory regulation or exceeds the permitted use, you will need to obtain permission directly from the copyright holder.

CHAPTER 2

Considering the Cross-Boundary Environmental and Social Implications of the EU's Carbon Dioxide Removal Strategy in Brazil

Joana Portugal-Pereira, Aline Cristina Soterroni, Antonella Mazzone, and Jiesper Strandsbjerg Tristan Pedersen

Policy Highlights To achieve the recommendation stated in the title, we propose the following:

- The European Commission should implement stricter regulations for domestic emission reduction and leverage EU Regulation on monitoring, verification, and reporting (MVR) of GHG emissions.

J. Portugal-Pereira (✉)
IN+ Centre for Innovation, Technology and Policy Research, Instituto Superior Técnico, Universidade de Lisboa, Lisboa, Portugal
e-mail: joana.portugal@tecnico.ulisboa.pt

A. C. Soterroni
University of Oxford, Oxford, UK
e-mail: aline.soterroni@biology.ox.ac.uk

© The Author(s) 2024
E. Galende Sánchez et al. (eds.), *Strengthening European Climate Policy*, https://doi.org/10.1007/978-3-031-72055-0_2

- The European Council should increase the budget for sustainable and regenerative agriculture and restorative forestry initiatives that enhance land-carbon sequestration.
- The European Council should revise the Multiannual Financial Framework (MFF) on novel CDR strategies and allocate funding to capacity building, and technology transfer to the Global South.
- The European Parliament should revisit the Renewable Energy Directive (RED) to establish safeguards and monitoring mechanisms to prevent negative impacts of BECCS projects in the Global South.
- Interdisciplinary collaborations for CDR strategies and governance should ensure they are not only technically viable but also socially acceptable, ethically sound, and culturally appropriate.

Keywords Climate neutrality · Carbon Dioxide Removal (CDR) · BECCS (Bioenergy with Carbon Capture and Storage) · Global South · Eucalyptus plantations

Introduction

The European Commission (EC) is committed to achieving climate neutrality through Carbon Dioxide Removal (CDR) strategies. However, the potential impacts on the Global South, particularly land ecosystems and local communities, are not well understood. CDR involves human-led initiatives to remove CO_2 from the atmosphere, such as natural sinks and/or capturing CO_2 during combustion. Conventional CDR strategies like reforestation, afforestation initiatives, and the use of novel methods, namely Bioenergy with Carbon Capture and Storage (BECCS) facilities are often used to mitigate climate change.

A. Mazzone
University of Bristol, Bristol, UK
e-mail: antonella.mazzone@ouce.ox.ac.uk

J. S. T. Pedersen
Utrecht University, Utrecht, The Netherlands
e-mail: jiespertristan@gmail.com

BECCS is a technology that produces energy from biomass, captures the CO_2 released during combustion, and then stores it underground. This process creates a net removal of CO_2 from the atmosphere making BECCS a promising tool for combating climate change (Butnar et al., 2020). According to the IPCC (2019), if the world continues in its current trend of intensive resource consumption, approximately 760 Mha of converted land, most likely located in the Global South, would be needed to compensate for high levels of remaining greenhouse gas (GHG) emissions.

In the Latin American and Caribbean regions, these dynamics worsen ecological and social inequalities, leading to conflicts stemming from resource extraction. Neo-colonialism, rooted in colonial times, refers to the disparities that persist in resource exploitation from Global South countries to power needs of colonial countries (Dorn, 2022; Nkrumah, 1965). Within postcolonial studies, energy transitions, and climate justice, green colonialism refers to the disparities that persist echoing historical colonial legacies within the context of resource exploitation.

Previous studies have addressed conflicts of energy transition agendas of the Global North to the Global South resources, such as renewables and hydrogen programmes, green extractivism, and subtler forms of carbon colonialism (Dorn, 2022; Zografos, 2022). This work proposes a mixed-methods approach to estimate the potential additional land required outside European boundaries, taking Brazil as a case, to meet the EU's Net Zero Emissions (NZE) goal and assess the environmental and social consequences beyond EU borders. It provides valuable insights for EU policymakers on the central differences in the meaning of energy justice and explores how philosophical notions of justice can help move away from unwanted green neo-colonialism agendas.

SCIENCE-BASED EVIDENCE FINDINGS

We developed a mixed-method approach combining modelling-based and social science qualitative methodological principles. Figure 2.1 depicts the key steps employed in this study.

The study began with an analysis of EC Green Deal policies, focusing on reducing GHG emissions by 2050 and conventional and novel CDR strategies within the EU. The next step involved assessing the additional land area needed in Brazil to offset European GHG emissions, using principles of net primary production and carbon capture by *Eucalyptus*

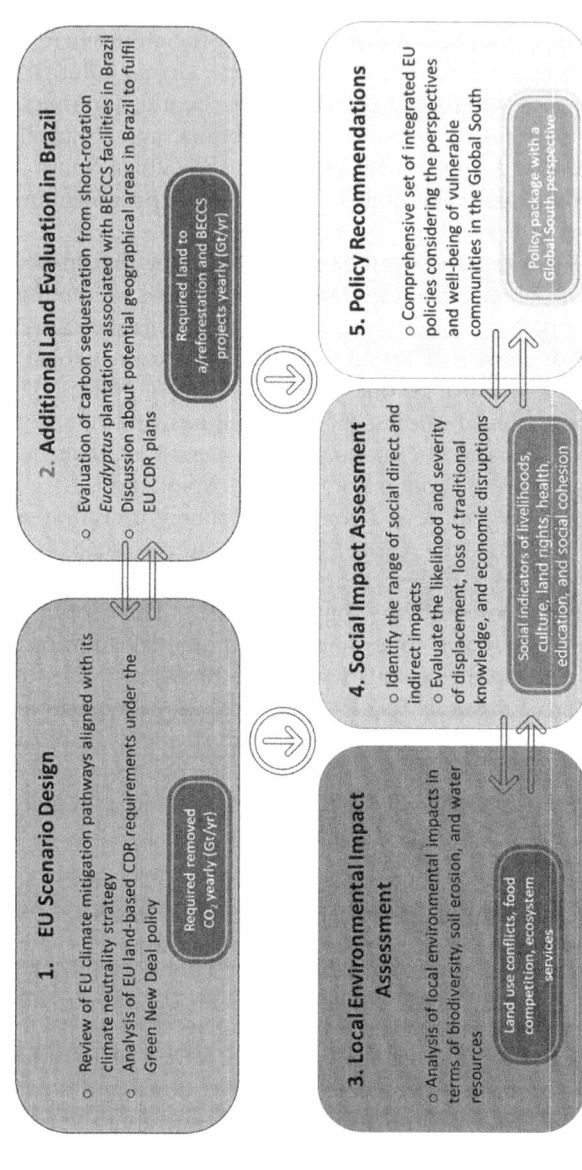

Fig. 2.1 Analytical framework of this study (*Source* Own elaboration)

plantations associated with bioenergy with BECCS facilities. A baseline scenario of the GLOBIOM-Brazil model was used to estimate potential available lands in Brazil by 2050. Socio-environmental impacts were evaluated using both qualitative and quantitative methods. Finally, concrete policy recommendations were formulated based on the findings of the study.

Land Requirements in Brazil to Accomplish the EC's NZE Goal

We explored two alternative pathways, distinguished by different levels of remaining emissions in the EU resulting from additional measures and the EU levels of net carbon sink removals from land-based CDR strategies from a/reforestation projects in the Land Use, Land-Use Change, and Forestry (LULUCF) sector and BECCS technologies. Projections suggest that even with additional measures implemented as part of the climate package (EEA, 2023), residual GHG emissions in the EU will still exist, reaching 2.48 and 1.74 $GtCO_2$ yearly by 2030 and 2050, respectively. Conversely, the European Scientific Board on Climate Change (ESABCC, 2024) estimates a capacity of net sink of the EU's LULUCF sector varying from -400 to -100$MtCO_2$ and a removal potential of BECCS technologies within EU ranging between –336 and -70$MtCO_2$ yearly by 2050.

To achieve climate neutrality by 2050, any remaining GHG within the EU that cannot be removed domestically will need to be offset elsewhere. This entails voluntary cooperation between the EU and third-party countries, as outlined in Article 6 of the Paris Agreement (UNFCCC, 2015) and detailed in the Paris rulebook (UNFCCC, 2021).

If this extra land would be met in Brazil, we considered the implementation of large-scale short-rotation *Eucalyptus* plantations in a monoculture regime with a harvest cycle of 7 years, a density of 2222 tree.ha^{-1} and a carbon content of 186 tC.ha^{-1}. *Eucalyptus* plantations are commonly used in afforestation projects due to their rapid growth rate, high biomass production, and efficient carbon sequestration capabilities, making them effective in offsetting CO_2 emissions (Portugal-Pereira et al., 2023). Furthermore, we considered BECCS projected to be coupled into thermal power plants with a capture efficiency of 90% and a carbon penalty of 24%.

Overall, we estimate that *Eucalyptus* plantation would remove 68.2 tonnes of CO_2 per hectare. If CDR strategies in the EU are speedily employed (low-risk case), the necessity for additional carbon dioxide removal beyond EU boundaries arises from 2031 after which it gradually increases, reaching, on yearly average for this period, 0.7 gigatonnes (Gt) of CO_2. This would result *in a cumulative removal of 14.7 $GtCO_2$ by 2050*. In the case of slow deployment of CDR strategies within EU territory, the removal starts as soon as in 2025, steadily increasing at an average rate of 0.9 $GtCO_2$ yearly for this period, with a cumulative removal of nearly *23.0 $GtCO_2$* during this period (high-risk case).

Under a low-risk case, the EU's land sinks potential and rapid deployment of climate change mitigation strategies could result in no additional land requirements until 2031, but by 2050, it would require 100.3 million hectares of land, equivalent to 1.2 times the total projected croplands in Brazil by 2050 (Soterroni et al., 2023).

If the LULUCF sinks are limited and the EU's CDR strategy is more conservative (high-risk case), annual land needs start in 2025, with 1.1 million hectares accumulating to 152.5 million hectares by 2050, which is nearly 80% of the entire estimated pasturelands in Brazil by 2050 under a baseline scenario (Soterroni et al., 2023). The allocation of this area in Brazil would intensify the competition for land within the country, as the agricultural abandonment by 2050 would provide only 71.0 million hectares that could be used for the EU's CDR strategy, according to the GLOBIOM-Brazil baseline scenario. It would also demand the restoration of up to 81.0 million hectares of degraded pastures on top of Brazil's existing commitments of pasture recovery under the national agricultural plan (ABC + Plan) and the Paris Agreement. The geographical location of these areas is expected to be located in the mid-central region of Brazil and the Southeast region of the Atlantic Forest biome (Soterroni et al., 2018, 2023). This understanding of the projected land distribution is crucial for sustainable land-use planning and resource allocation.

Unintended Consequences: Environmental and Social Threats

The deployment of short-rotation *Eucalyptus* plantations in monoculture regimes poses significant threats to terrestrial ecosystem services, including food security, freshwater, soil resources, and air and water quality regulation. Factors such as land governance, management regimes, edaphoclimatic conditions, competing land demands, and deployment scale can influence these risks (Calvin et al., 2021; Humpenöder et al., 2014; Portugal-Pereira et al., 2016) and may have unknown ecosystem consequences (DeFries et al., 2004). Such, rather local, aspects are to a lesser degree considered in models.

Intensive management of *Eucalyptus* forests promotes soil compaction, infertility, and erosion (Landis et al., 2018). Additionally, the efficacy of soil carbon capture (SOC) can be reversible if continuous management practices to enhance soil carbon are not rigorously upheld (Andren & Katterer, 2001). The cultivation of *Eucalyptus* plantations has the potential to displace food production, resulting in heightened food prices (Smith et al., 2019). Concerns also arise regarding the potential pressure on deforestation despite explicit prohibitions on direct illegal deforestation (Ferrante & Fearnside, 2020).

Large-scale biomass plantations also affect water resources, exacerbating water scarcity in regions already under pressure, particularly in irrigated systems (Heidari et al., 2021). This is particularly serious in the Northeast region of Brazil. Monoculture production also affects biodiversity, especially when natural landscapes are converted into monoculture plantations or peatlands are drained (IPBES, 2019).

Regarding social risks, indigenous and local communities are susceptible to negative impacts stemming from land-based CDR strategies centred on afforestation, particularly those characterised by large monocultures of non-native species. Previous experiences with afforestation programmes for the development and internationalisation of the pulp industry show there are direct and indirect negative consequences of such programmes for local communities. Direct impacts encompass potential land dispossession, shifts in customary livelihoods, modifications in soil composition with concomitant implications for food security, and health-related alterations arising from soil contamination. Moreover, direct detriments extend to water scarcity and the depletion of cultural sites. A remarkable example is the case of Aracruz, a former Norwegian–Brazilian cellulose company in the state of Espirito Santo (Kenfield, 2007).

Indirect deleterious consequences commonly manifest when afforestation initiatives are implemented in the proximate vicinities of indigenous land tenure. Illustratively, an ethnographic investigation conducted within the context of the Extractive Reserve (RESEX) Guai (Sapulcaia, Bahia State) reveals that local communities contiguous to *Eucalyptus* monocultures encountered water contamination with a consequent disruption in their traditional livelihood (in this case fishing). Another indirect negative consequence for local communities can be represented by the increase in land prices. Kröger (2012) reports an increase in land prices in areas designated for pulp projects, turning family farming economically unfeasible.

Considering these aforementioned socio-environmental consequences, integrated modelling exercises that estimate CDR strategies capacity in the Global South should consider socio-economic barriers and internalise social and environmental impacts into their least-cost optimised pathways. These could be implemented, for instance, by quantifying social and environmental externalities of large-scale conventional and novel CDR strategies that would add up the levelised abatement costs of CO_2.

Conclusions and Recommendations

After conducting our analysis, we present five science-based policy recommendations aimed at mitigating potential unintended consequences in the Global South:

The EC is urged to implement stricter climate action measures to reduce emissions within its territory. This includes revising member states' mitigation targets across key sectors like energy, transport, agriculture, and industry. These targets should be phased in over time, with yearly reviews based on technological advancements. The EU ETS, the existing carbon pricing mechanism, should be expanded with a carbon tax to incentivise emission reductions across sectors. This proposal is part of internal EU policy and regulations, involving both the Commission and Council.

The EC must champion nature-based solutions, such as regenerative agriculture, agroforestry, and restoration of ecosystems to increase land-carbon sequestration and conserve biodiversity and natural habits in the EU. This requires a reassessment of the LULUCF regulation, with stricter targets to enhance carbon sequestration in forests and other natural ecosystems. Financial incentives for landowners are crucial to ensure compliance with best practice guidelines. EC agencies, such as the European Environmental Agency (EEA), should collaborate with local authorities to provide effective land-carbon management practices while safeguarding local ecosystems and communities.

The European Council is urged to increase investments in R&D for novel CDR strategies and to facilitate technology transfer between the Global South and North regions. This entails revising the upcoming Multiannual Financial Framework (MFF) (EC, 2023) to incorporate a dedicated budget line for R&D on innovative CDR strategies, such as

BECCS, and allocate funds for technology transfer to/from the Global South. Furthermore, the Directorate-General for Research and Innovation should facilitate a dedicated Horizon Europe programme to develop novel CDR strategies. This programme should involve collaboration between EU and Global South research centres to drive innovation and address societal challenges, e.g., impact on local societies.

The European Commission must promote responsible carbon offsetting by revisiting the Renewable Energy Directive (RED) and establishing socio-environmental safeguards to mitigate and prevent negative impacts of BECCS projects on local ecosystems and communities in the Global South. Funding mechanisms within the EU ETS framework should be established for compensation schemes for local communities impacted by carbon offset projects.

The European Commission (EC) should leverage the existing framework provided by the EU Regulation on MVR of GHG emissions (EC, 2018) to encompass all sectors of the EU economy. This approach aims to enhance the accountability of carbon dioxide fluxes, ensuring both high credibility and the durability/permanence of removal efforts. This requires establishing rigorous standards for data collection, reporting methodologies, and independent audits. Key agencies to deliver these efforts include the EEA, which could offer technical expertise and support for the development and implementation of standardised methodologies; the Directorate-General for Climate Action (DG CLIMA), which could lead efforts to expand the scope of the MVR-GHG regulation and strengthen standards for data collection and reporting; and, the Joint Research Centre (JRC) that is well-positioned to develop scientifically robust methodologies.

This chapter explored a space for interdisciplinary knowledge exchange and co-creation, potentially leading to new research avenues and a more robust understanding of complex climate challenges between national models and local complexity. In opposition to a focus on overall techno-economic strategies (modelling, CDR methods, quantitative trade-offs), the SSH focused on local elements related to social impacts, species composition, and their connection with national policies. Although this created challenges related to scale, and methodologies, discussing these differences in focus and methods led to integrating local complexity into the national models, i.e., calibrating new variables into the model. As such, we managed to bridge a gap in climate change science, often addressed by social scientists.

Acknowledgements The authors express gratitude to Gerd Angelkorte for his valuable advice on land-use change dynamics in Brazil. J. Portugal-Pereira thanks the support of Brazilian National Council for Scientific and Technological Development (CNPq) (grant:307234/2020 307234/2020 7). J. Pedersen would like to acknowledge the financial support provided by the Portuguese Fundação para a Ciência e a Tecnologia (FCT) I.P./MCTES through national funds (PIDDAC)—UIDB/50019/2020 (https://doi.org/https://doi.org/10.54499/UIDB/50019/2020).

References

Andren, O., & Katterer, T. (2001). *Basic principles for soil carbon sequestration and calculating dynamic country-level balances including future scenarios* (J. Kimble, R. Follett, & B. Stewart, Eds.). Lewis Publishers.

Butnar, I., Li, P. H., Strachan, N., Portugal Pereira, J., Gambhir, A., & Smith, P. (2020). A deep dive into the modelling assumptions for biomass with carbon capture and storage (BECCS): A transparency exercise. *Environmental Research Letters, 15*(8), 084008.

Calvin, K., Cowie, A., Berndes, G., Arneth, A., Cherubini, F., Portugal-Pereira, J., ... & Smith, P. (2021). Bioenergy for climate change mitigation: Scale and sustainability. *GCB Bioenergy, 13*(9), gcbb.12863.

DeFries, R. S., Foley, J. A., & Asner, G. P. (2004). Land-use choices: Balancing human needs and ecosystem function. *Frontiers in Ecology and the Environment, 2,* 249–257.

Dorn, F. M. (2022). *Green colonialism in Latin America? Towards a new research agenda for the global energy transition. European Review of Latin American and Caribbean Studies, 114,* 137–146

EC. (2018). Regulation 2018/2066 of 19 December 2018 on the monitoring and reporting of greenhouse gas emissions pursuant to Directive 2003/87/EC of the European Parliament and of the Council and amending Commission Regulation (EU) No 601/2012. Brussels, Belgium.

EC. (2023). Commission communication "Mid-term revision of the MFF 2021–2027" European Commission, Brussels, Belgium.

EEA. (2023). *Total net greenhouse gas emission trends and projections in Europe.* European Environmental Agency. Copenhagen, Denmark.

ESABCC. (2024). *Towards EU climate neutrality Progress, policy gaps and opportunities.* European Scientific Advisory Board on Climate Change.

Ferrante, L., & Fearnside, P. M. (2020). The Amazon: Biofuels plan will drive deforestation. *Nature, 577*(7789), 170–170.

Heidari, A., Watkins Jr, D., Mayer, A., Propato, T., Verón, S., & De Abelleyra, D. (2021). Spatially variable hydrologic impact and biomass production tradeoffs associated with Eucalyptus (E. grandis) cultivation for biofuel production in Entre Rios, Argentina. *GCB Bioenergy, 13*(5), 823–837.

Humpenöder, F., Popp, A., Dietrich, J. P., Klein, D., Lotze-Campen, H., Bonsch, M., ... & Müller, C. (2014). Investigating afforestation and bioenergy CCS as climate change mitigation strategies. *Environmental Research Letters, 9*(6), 064029.

IPBES. (2019). *Global assessment report on biodiversity and ecosystem services of the Intergovernmental Science-Policy Platform on Biodiversity and Ecosystem Services* (E. Brondizio, J. Settele, S. Díaz, & H. Ngo, Eds.). IPBES Secretariat.

IPCC. (2019). Summary for policymakers. In *Special report on climate change, desertification, land degradation, sustainable land management, food security, and greenhouse gas fluxes in terrestrial ecosystems*.

Kenfield, I. (2007). Taking on big cellulose: Brazilian indigenous communities reclaim their land. *NACLA Report on the Americas, 40*(6), 9–13.

Kröger, M. (2012). The expansion of industrial tree plantations and dispossession in Brazil. *Development and Change, 43*(4), 947–973.

Landis, D. A., Gratton, C., Jackson, R. D., Gross, K. L., Duncan, D. S., Liang, C., Meehan, T. D., Robertson, B. A., Schmidt, T. M., Stahlheber, K. A., Tiedje, J. M., & Werling, B. P. (2018). Biomass and biofuel crop effects on biodiversity and ecosystem services in the North Central US. *Biomass and Bioenergy, 114*, 18–29.

Nkrumah, K. (1965). *Neo-colonialism: The last stage of imperialism*. Thomas Nelson & Sons, Ltd.

Portugal-Pereira, J., Köberle, A. C., Soria, R., Lucena, A. F. P., Szklo, A., & Schaeffer R. (2016). Overlooked impacts of electricity expansion optimisation modelling: The life cycle side of the story. *Energy, 115, Part 2*, 1424–1435. https://doi.org/10.1016/j.energy.2016.03.062

Portugal-Pereira, J., Carvalho, F., Rathmann, R., Szklo, A., Rochedo, P., & Schaeffer R. (2023). *Ch18: The potential of biomass. In Handbook on the geopolitics of the energy transition*. Elgar.

Smith, P., Nkem, J., Calvin, K., Campbell, D., Cherubini, F., Grassi, G., ... & Taboada, M. A. (2019). Interlinkages between desertification, land degradation, food security and GHG fluxes: Synergies, trade-offs and Integrated Response Options. In *Special report on climate change, desertification, land degradation, sustainable land management, food security, and greenhouse gas fluxes in terrestrial ecosystems*. Cambridge University Press.

Soterroni, A. C., Império, M., Scarabello, M. C., Seddon, N., Obersteiner, M., Rochedo, P. R., ... & Alencar, A. A. (2023). Nature-based solutions are critical for putting Brazil on track towards net-zero emissions by 2050. *Global Change Biology, 29*(24), 7085–7101.

Soterroni, A. C., Mosnier, A., Carvalho, A. X., Câmara, G., Obersteiner, M., Andrade, P. R., ... & Ramos, F. M. (2018). Future environmental and agricultural impacts of Brazil's Forest Code. *Environmental Research Letters, 13*(7), 074021.

UNFCCC. (2015). Paris Agreement. *Conference of the Parties, 21932*(December), 32.

UNFCCC. (2021). Glasgow Climate Pact, decision 1/CMA.3. Conference of the Parties.

Zografos, C. (2022). The contradictions of Green New Deals: Green sacrifice and colonialism. *Soundings, 80*(80), 37–50.

Open Access This chapter is licensed under the terms of the Creative Commons Attribution 4.0 International License (http://creativecommons.org/licenses/by/4.0/), which permits use, sharing, adaptation, distribution and reproduction in any medium or format, as long as you give appropriate credit to the original author(s) and the source, provide a link to the Creative Commons license and indicate if changes were made.

The images or other third party material in this chapter are included in the chapter's Creative Commons license, unless indicated otherwise in a credit line to the material. If material is not included in the chapter's Creative Commons license and your intended use is not permitted by statutory regulation or exceeds the permitted use, you will need to obtain permission directly from the copyright holder.

CHAPTER 3

Weaving a Transformative Circular Textile Policy Through a Socio-Environmental Justice Lens

Lis J. Suarez-Visbal, Martin Calisto Friant, Anna Härri, Veerle Vermeyen, Abe Hendriks, Blanca Corona Bellostas, and Jesus Rosales Carreon ⓘ

Policy Highlights To achieve the recommendation stated in the title, we propose the following:

- Tackle overproduction and overconsumption in the EU Strategy for Sustainable and Circular Textiles.

L. J. Suarez-Visbal · V. Vermeyen · A. Hendriks · B. Corona Bellostas
Utrecht University, Utrecht, The Netherlands
e-mail: l.j.suarezvisbal@uu.nl

V. Vermeyen
e-mail: Veerle.vermeyen@kuleuven.be

A. Hendriks
e-mail: a.hendriks@uu.nl

B. Corona Bellostas
e-mail: b.c.coronabellostas@uu.nl

© The Author(s) 2024
E. Galende Sánchez et al. (eds.), *Strengthening European Climate Policy*, https://doi.org/10.1007/978-3-031-72055-0_3

- Address the impacts of EU Circular Economy Textile policies on the Global South from both SSH and STEM perspectives to ensure positive social and environmental outcomes.
- Make Just Transition policies globally accountable and alleviation mechanisms integral to the Textile Strategy rather than supplementary corrective measures.
- Include meaningful participatory mechanisms that ensure the democratic inclusion of different voices and actors.
- Reverse the burden of proof and provide educational, financial, and legal assistance accounting for multiple vulnerabilities (e.g., gender or type of worker).

Keywords Circular Economy (CE) · Planetary boundaries · Environmental justice · EU textile strategy · Sustainability policies

Introduction

The way we produce and consume has a significant impact on both the environment and society. We are already overshooting six of the nine planetary boundaries (Richardson et al., 2023), risking irreversible environmental degradation and jeopardising the well-being of current and future generations. To address this, the Circular Economy (CE) concept is

---

M. Calisto Friant
Universitat Autònoma de Barcelona, Bellaterra, Spain
e-mail: martin.calisto@uab.cat

A. Härri
LUT University, Lappeenranta, Finland
e-mail: anna.harri@lut.fi

V. Vermeyen
KU Leuven, Leuven, Belgium

J. R. Carreon (✉)
Copernicus Institute of Sustainable Development, Utrecht University, Utrecht, The Netherlands
e-mail: j.rosales.carreon@uu.nl

becoming a crucial narrative guiding international, national, and sectoral sustainability policies.

As one of the world's largest economies, the EU significantly influences global environmental and social conditions. One critical sector targeted by the European Green Deal is textiles, given that this sector ranks fourth for the highest impact on the environment and climate change, the third highest for water and land use, and fifth for primary raw materials (EEA, 2022). Additionally, from extraction to end-of-life, the textile sector remains labour-intensive, providing millions of jobs to workers in Europe and the Global South, where most textiles are produced and most textile waste is exported (Köhler et al., 2021). Furthermore, the sector is highly feminised, as women are overrepresented in the lowest-paying jobs (Fletcher & Tham, 2014).

The European Commission has developed a new EU strategy for sustainable and circular textiles to address these critical sustainability challenges. It aims to harmonise the European Green Deal, the Circular Economy Action Plan, and the European industrial strategy to develop a greener, more competitive textile sector. Despite these ambitious plans, there is a lack of research on the socio-ecological implications of these policies from a social and environmental justice perspective. This study addresses this research gap by answering the following question:

How Can EU Textile Policies Enable the Transition to a Fair and Sustainable Circular Society?

To answer this question, we analysed the EU Textile Strategy and 9 of the 25 actions in its annex, namely the directives, regulations, or communications with direct policy relevance. The remaining actions were either still in development and thus unavailable or were reports and working documents that did not evidence the EU's current policy. Further, we included the Just Transition Fund (Regulation (EU) 2021/1056), even though it is not part of the annex of the EU Textile Strategy. We analyse in total 11 EU policy documents (see Appendix 1).

The insights from our analysis are particularly relevant for policy-makers at the European Commission and researchers interested in CE governance.

The analysis consisted of three steps:

1. **Literature review** of policy research on the circular transition in the textile industry in the SSH and STEM fields
2. **Co-development of an interdisciplinary analytical framework** based on environmental justice, CE, sustainability, and post-growth literature applied to the textile sector: The framework was developed during a co-production workshop, where all chapter authors combined the insights from their specific SSH and/or STEM fields in a comprehensive interdisciplinary approach. The framework considers four justice dimensions, namely recognitive, distributive, procedural, and restorative, as well as environmental boundaries. For each dimension, a set of questions was created to evaluate the policies, which can be found in Appendix 2.
3. **Application of the interdisciplinary framework** to the chosen policies: Each policy's findings related to each dimension were summarised in an Excel sheet and colour-coded based on adequacy. Through an iterative process, multiple authors collaborated to collectively analyse and discuss the policy documents (see Appendix 3).

Analytical Framework

Our interdisciplinary analytical framework contributes to developing a transformative and just CE transition in the textile value chain. A transformative CE could become a driver for systemic change away from present unsustainable production and consumption structures. A technical shift and a shift in values, behaviours, and institutional structures are needed to create a more regenerative, democratic, and equitable system. A transformative CE should empower the most vulnerable people and allow all humans to shape their society and future equitably (Calisto Friant et al., 2023; Suarez-Visbal et al., 2024).

Academics studying social and environmental justice have identified four core dimensions, namely recognitive, distributive, procedural, and restorative justice (Abram et al., 2022).

Recognitive justice is about who is recognised in a socio-ecological transition. It seeks to ensure that the views of the most marginalised people are heard and recognised (Parsons et al., 2021). Distributive justice looks at the distribution of benefits and harms resulting from socio-ecological change. Procedural justice asks what procedures can prevent the creation or reproduction of injustice in transitions or bring justice to harmed communities. Restorative justice is about repairing harm caused by specific behaviour both on the social and ecological sides (McCauley & Heffron, 2018).

We added a fifth dimension to reflect the environmental boundaries of the Earth and the key role that more-than-human nature plays in a just transition (Sharpe et al., 2023). The dimension recognises that human beings are an integral part of nature and that nature is a subject of rights. A just and sustainable CE transition occurs when all five socio-ecological justice and sustainability dimensions are integrated, revealing multiple tensions and trade-offs that must be addressed and negotiated.

Analysis of Relevant EU Policies

This section describes the main insights from analysing the 11 selected EU policies (Fig. 3.1).

Environmental Dimension

The environmental dimension is present through initiatives to promote greater transparency to consumers regarding the impact of production and consumption or through the promotion of strategies on the value-retention hierarchy. However, higher value strategies such as refuse and reduce are largely missing.

Furthermore, no policies address the root causes of negative socio-ecological impacts, like overproduction, overconsumption, and excessive advertising (Sharpe et al., 2023). Although the EU CE Textile Strategy acknowledges that textiles are overproduced and overconsumed, none of the policies underlying the strategy address this.

While some policies mention planetary boundaries, none recognise that the EU lives beyond its fair share of planetary resources. The policies lack clear targets and limits to reduce overall environmental impacts, so the EU's footprint would fall within sustainable planetary boundaries.

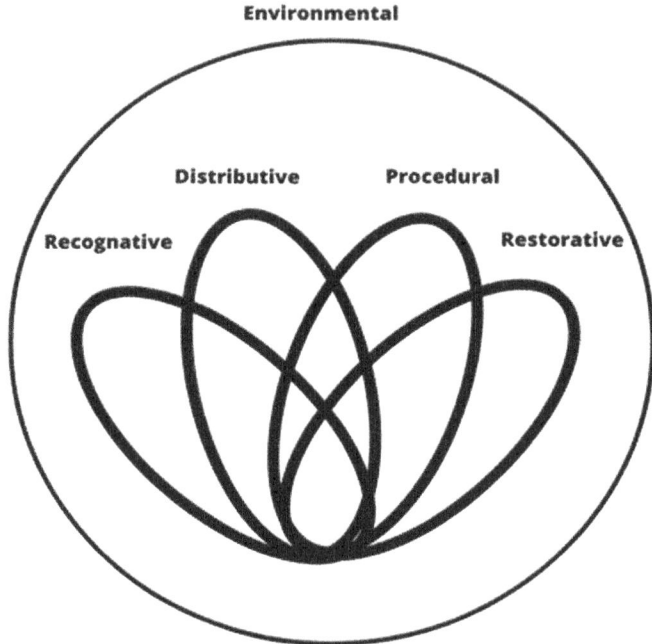

Fig. 3.1 Interdisciplinary analytical framework combining five interrelated and interconnected dimensions of socio-environmental justice (*Source* Modified from Härri and Levänen [2024, 14]

Recognitive Dimension

Together, the policies will have a considerable impact on various stakeholders outside of the EU, as the majority of clothes sold in the EU are made in the Global South. Unfortunately, this is not adequately recognised in the policies.

Many people in the Global South are in vulnerable positions and lack resilience, especially workers in low-paying sectors (such as women and migrants) and people with limited access to education or upskilling. Some actors hold simultaneous vulnerabilities, such as being female, a refugee, or an undocumented waste picker. Furthermore, these vulnerabilities could be exacerbated depending on each country's labour security structure (Suarez-Visbal et al., 2024).

The Due Diligence Directive is the most inclusive of the analysed policies. It recognises the vulnerabilities of several actors but could benefit from explicitly recognising vulnerable countries, informal workers, ethnic minorities, and agricultural workers. These groups lack adequate recognition across all analysed policies. Informal workers are critical and need to be included as they are likely the most affected by e.g., the relocalisation of the industry to the Global North. Moreover, more provisions should include women across the different policies. In some policies, they are simply mentioned, but no tangible actions are taken to address their vulnerability.

Furthermore, most policies exclude Small and Medium-sized Enterprises (SMEs) from responsibilities and risk the significant impacts caused by SMEs remaining unaddressed. Instead of exclusion, financial and technical support is needed to help them comply. The policies fail to account for the multiplicity of vulnerabilities and the multidimensional and contextual nature of discrimination, exploitation, and alienation. They provide an EU-centric idea of sustainable textiles that lacks alternative visions of the future, especially from affected peoples and ecosystems in the Global South.

Distributive Dimension

The policies have an unfair distribution of costs and benefits that will likely disproportionally affect the most vulnerable. Current policies focus mainly on severe social and environmental impacts (like forced labour) but do not challenge unsustainable business models and purchasing practices that foster social and environmental impacts in the value chain.

Although quality jobs are mentioned in 8 of the 11 policy documents, there is no comprehensive definition of what this entails, especially for workers outside the EU. Such a definition should include living wages, well-being, work-life balance, non-discrimination, collective bargaining, and inclusiveness.

The amendments to the Waste Directive underestimate the significant impact of implementing these policies on the Global South. EU policies risk negatively affecting their livelihood by not including SMEs or adequately recognising marginalised workers and textile waste pickers in the Global South. As textile waste currently ends up in the Global South, the eco-modulation of the Extended Producer Responsibility (EPR) fee should be increased to allow for the full and sustainable recovery of these wastes, regardless of where they end up (Thapa et al., 2022). If textiles are exported for reuse or recovery in the Global South, they should be accompanied by financial and technical resources to ensure a second life or proper disposal.

The Green Deal considers the Just Transition Mechanism (JTM) as a means to address the asymmetries of the zero carbon and circular transition. However, the EU Textile Strategy does not consider it an integral action. An extended globally accountable JTM could be incorporated into the policy package of the EU textile strategy. Additionally, a harmonised classification of "green activities" enabled through the "European taxonomy" should include environmental and social criteria in their evaluation, as the Trade Union Confederation (ETUC) and several NGOs have already expressed.

Procedural Dimension

Our policy analysis revealed that throughout the 11 policy documents, there are very few participatory mechanisms to ensure that different actors are democratically included and given decision-making power in their transposition and implementation. There are also very few education and empowerment mechanisms that encourage greater understanding and participation of all voices.

Some policies establish consultation bodies, such as the eco-design forum of the Eco-design Regulation and the Committee established by the Directive on Empowering Consumers for the Green Transition. Moreover, some policies require third actors to carry out participatory mechanisms, such as the Due Diligence Directive, which asks companies to ensure stakeholder engagement while carrying out their due diligence duties. However, these participatory mechanisms are poorly described and defined in these policies. They may become simple ticking-the-box exercises and consultations that give very little tangible power for affected people to shape the decisions that affect them. The current policies are thus unlikely to ensure that the most marginalised voices are truly heard and have decision-making power.

Restorative Dimension

There is little to no support for the Global South to adapt to the multiple new provisions set by the policies of the textile strategy. Yet these policies will likely significantly impact the most vulnerable suppliers, farmers, and formal and informal workers in the Global South, who will have to change their production practices and fulfil new administrative requirements. In addition to this, EU policies do not sufficiently encourage companies to transform their unsustainable purchasing practices and business models. Fast fashion strategies paying very low prices to producers and often forcing workers to overwork to fulfil short-term orders will thus likely continue. In these conditions, there will probably be little real transformation in textile value chains as the root causes of these socio-ecological problems are not addressed.

Furthermore, the legal liability for companies that commit social and environmental impacts is weak in the Due Diligence Directive because there are few tangible mechanisms to ensure access to justice for affected people, and there is no reversal of the burden of proof (affected people have to prove that a company has violated their human and environmental rights). This is especially problematic for the most vulnerable, marginalised, and discriminated people in the Global South, who often lack the knowledge, awareness, and financial resources to uphold their social and environmental rights.

Conclusion and Recommendations

The 11 analysed policy instruments are an ambitious first step of policy commitment towards circularity. However, more emphasis must be placed on the scope, breadth, and depth of the actions proposed to generate tangible socio-ecological changes in the textile sector. This could be achieved with a transformative CE lens that redresses imbalances in the linear economic system, ensuring these legislations are truly fair, democratic, and sustainable. Though bridging the methodologies of STEM and SSH disciplines can pose communication barriers, the diverse perspectives ultimately cultivate synergies that drive progress and address complex challenges with depth and breadth.

More specifically:

The EU policies should acknowledge that current overproduction and overconsumption patterns are the root causes of current socio-ecological problems. The EU should establish clear targets and limits to reduce textile overproduction and ensure its ecological footprint stays within planetary boundaries.

There should be an increased recognition of the multiple lived realities and vulnerabilities of stakeholders affected by EU legislation (both within and outside the EU). This will help ensure that EU policies truly recognise the implications of their policies on marginalised, discriminated, and exploited people and establish sufficient mitigation measures to address them. A way to address this is by collecting disaggregated data on multiple vulnerabilities (e.g., gender, type of worker, etc.).

Since the current JTM fails to consider the global dimension, a globally accountable JTM should be incorporated into the policy package of the EU textile strategy. Moreover, the policies of the strategy are not sufficiently discouraging unsustainable business models and fast fashion—the key drivers of socio-ecological problems in the sector (Sharpe et al., 2023). It is thus key for the EU to provide greater financial, technical, and technological assistance and support for suppliers and countries in the Global South to transform their production practices sustainably.

Moreover, textile EPRs should have global accountability to cover all regions where EU textile waste is currently exported. This will ensure a collective Global responsibility to sustainably handle textile waste, regardless of where it ends up.

We found a general lack of sufficiently diverse stakeholder engagement and participation throughout all policies, especially including perspectives from marginalised people. EU policies should include meaningful participatory mechanisms to ensure that different voices are democratically included and given decision-making power (such as citizen assemblies and greater engagement with vulnerable people in the textile value chain). Such mechanisms should minimise asymmetries in power between different stakeholders and ensure the empowerment of the voices, interests, and visions of the most marginalised peoples from the Global North and South alike.

There is a general lack of access to justice and reparation for socio-environmental harm caused by EU companies in the Global South. To address this, EU policies should reverse the burden of proof so that affected stakeholders do not face unsurmountable legal challenges to prove their case. Moreover, the EU should provide educational, financial, and legal assistance to ensure access to justice for affected people.

Appendixes

Appendix 1

Suarez-Visbal, L., Calisto Friant, M., Corona Bellastos, B., Rosales Carreon, J., Härri, A., Vermeyen, V., & Hendriks, A. (2024). *List of 11 analysed policy documents*. Zenodo. https://zenodo.org/records/10783492.

Appendix 2

Suarez-Visbal, L., Calisto Friant, M., Corona Bellastos, B., Rosales Carreon, J., Härri, A., Vermeyen, V., & Hendriks, A. (2024). *Interdisciplinary analytical framework on the socio-ecological justice and sustainability implications of a circularity transition*. Zenodo. https://zenodo.org/records/10847421.

Appendix 3

Suarez-Visbal, L., Calisto Friant, M., Härri, A., Vermeyen, V., Hendriks, A., Corona Bellastos, B., & Rosales Carreon, J. (2024) *Results of the interdisciplinary policy analysis*. Zenodo. https://zenodo.org/records/10839063.

References

Abram, S., Atkins, E., Dietzel, A., Jenkins, K., Kiamba, L., Kirshner, J., Kreienkamp, J., Parkhill, K., Pegram, T., & Santos Ayllón, L. M. (2022). Just transition: A whole-systems approach to decarbonisation. *Climate Policy, 22*(8), 1033–1049.

Calisto Friant, M., Vermeulen, W. J. V., & Salomone, R. (2023). Transition to a sustainable circular society: More than just resource efficiency. *Circular Economy and Sustainability, 4*, 23–42.

EEA. (2022). *Textiles and the environment: The role of design in Europe's circular economy—European Environment Agency* [Briefing]. https://www.eea.europa.eu/publications/textiles-and-the-environment-the

Fletcher, K., & Tham, M. (Eds.). (2014). *Routledge handbook of sustainability and fashion*. Routledge.

Härri, A., & Levänen, J. (2024). "It should be much faster fashion"—Textile industry stakeholders' perceptions of a just circular transition in Tamil Nadu, India. *Discover Sustainability, 5*(1), 1–19.

Köhler, A., Watson, D., Trzepacz, S., Löw, C., Liu, R., Danneck, J., Konstantas, A., Donatello, S., & Faraca, G. (2021). *Circular economy perspectives in the EU textile sector*. Publications Office of the European Union.

McCauley, D., & Heffron, R. (2018). Just transition: Integrating climate, energy and environmental justice. *Energy Policy, 119*, 1–7.

Parsons, M., Fisher, K., & Crease, R. P. (2021). Environmental justice and indigenous environmental justice. In M. Parsons, K. Fisher, & R. P. Crease (Eds.), *Decolonising blue spaces in the anthropocene: Freshwater management in Aotearoa New Zealand* (pp. 39–73). Springer International Publishing.

Richardson, K., Steffen, W., Lucht, W., Bendtsen, J., Cornell, S. E., Donges, J. F., Drüke, M., Fetzer, I., Bala, G., von Bloh, W., Feulner, G., Fiedler, S., Gerten, D., Gleeson, T., Hofmann, M., Huiskamp, W., Kummu, M., Mohan, C., Nogués-Bravo, D., ... Rockström, J. (2023). Earth beyond six of nine planetary boundaries. *Science Advances, 9*(37), eadh2458.

Sharpe, S., Retamal, M., & Brydges, T. (2023). Beyond growth: A well-being economy for the textile and garment sector. *Public Health Research & Practice, 33*(2).

Suarez-Visbal, L. J., Carreón, J. R., Corona, B., Hoffman, J., & Worrell, E. (2024). Transformative circular futures in the textile and apparel value chain: Guiding policy and business recommendations in the Netherlands, Spain, and India. *Journal of Cleaner Production, 447*.

Thapa, K., Vermeulen, W. J. V., Deutz, P., & Olayide, O. (2022). Ultimate producer responsibility for e-waste management–A proposal for just transition in the circular economy based on the case of used European electronic equipment exported to Nigeria. *Business Strategy and Development*.

Open Access This chapter is licensed under the terms of the Creative Commons Attribution 4.0 International License (http://creativecommons.org/licenses/by/4.0/), which permits use, sharing, adaptation, distribution and reproduction in any medium or format, as long as you give appropriate credit to the original author(s) and the source, provide a link to the Creative Commons license and indicate if changes were made.

The images or other third party material in this chapter are included in the chapter's Creative Commons license, unless indicated otherwise in a credit line to the material. If material is not included in the chapter's Creative Commons license and your intended use is not permitted by statutory regulation or exceeds the permitted use, you will need to obtain permission directly from the copyright holder.

CHAPTER 4

Adapting to Heatwaves: Reframing, Understanding, and Translating Strategies from India to the EU

Laura Menatti, *Anna-Katharina Brenner*,
Joyshree Chanam, *Marina Knickel*, *Hari Sridhar*,
and Corey Bunce

Policy Highlights To achieve the recommendation stated in the title, we propose the following:

- Adaptation should be reframed as situated, relational, and long-term processes within socio-ecological systems to maximise the value of nature-oriented, place-based, and community-driven strategies.

Laura Menatti and Corey Bunce contributed equally to this work.

L. Menatti (✉) · M. Knickel · H. Sridhar · C. Bunce
Konrad Lorenz Institute, Klosterneuburg, Austria
e-mail: laura.menatti@gmail.com

M. Knickel
e-mail: marina.knickel@kli.ac.at

H. Sridhar
e-mail: harisridhar1982@gmail.com

© The Author(s) 2024
E. Galende Sánchez et al. (eds.), *Strengthening European Climate Policy*, https://doi.org/10.1007/978-3-031-72055-0_4

- Interdisciplinary approaches integrating STEM and SSH perspectives provide better understanding of adaptation strategies and lead to innovative knowledge for guiding adaptation actions.
- Adaptation policies in the EU should draw upon the Global South given their long history of adaptation to extreme temperatures.
- Translating learning possibilities requires intercultural and intergeographical tailoring of policy according to local options, needs, and resources, and not just mere knowledge transfer.
- To foster mutually beneficial global interrelations, the EU should emphasise opportunities for social and epistemic justice, and interconnectedness.

Keywords Adaptation · Interdisciplinarity · Heatwaves · Vulnerability · Mutual Learning

Introduction

According to the European Climate Risk Assessment (EUCRA), Europe is the fastest-warming continent, with heatwaves posing threats to ecosystems, infrastructures, and health at various socio-geographical levels (EEA, 2024). Despite increasing attempts (e.g., https://climate-adapt.eea.europa.eu/en/news-archive), EUCRA finds that European policies fail to cope adequately with growing climate risks, calling for urgent actions. Similar concerns are echoed in IPCC (2023), highlighting adaptation gaps like unequal access to cooling solutions and short-term

---

C. Bunce
e-mail: corey.bunce@kli.ac.at

A.-K. Brenner
Leibniz Institute of Ecological Urban and Regional Development (IÖR), Dresden, Germany
e-mail: a.brenner@ioer.de

J. Chanam
Konrad Lorenz Institute, Klosterneuburg, Austria
e-mail: joyshreechanam@gmail.com

fragmented interventions. To better understand risks and adaptation strategies, EUCRA advocates for an integrated policy approach, focusing on vulnerable groups (e.g., elderly, gender-sensitive, those in poorly built dwellings) and regions, like Southern Europe. The document emphasises that reducing climate risk goes beyond traditional health policies, citing spatial planning and infrastructure management as keys. Our chapter builds on this, offering insights for future EU policies with a focus on cooperation and social justice (EEA, 2024, 33). We innovatively suggest broader cooperation beyond Europe, especially with areas experienced in heat adaptation in the Global South (Nagendra, 2018).

In this chapter, we first propose a reframing of adaptation oriented towards identification, analysis, and translation of strategies by drawing on our multiple disciplinary perspectives. Second, we use this interdisciplinary framing to better understand heatwave responses in India as examples of adaptation to climate change. We pay specific attention to nature-oriented, place-based, and community-driven strategies applied to vulnerable and deprived urban areas. Third, based on the examples, we present guiding principles for translating adaptation strategies between the Global South and the EU which centre mutual learning, interconnectedness, and both social and epistemic (concerning knowledge and understanding) justice.

Interdisciplinary Methodology

Across these three phases, we systematically used an interdisciplinary approach, aimed at linking and integrating ideas, concepts, methods, and research practices. Knowledge integration was fundamental, as we co-developed an interdisciplinary framework for adaptation during several dedicated events. Specifically, we performed a conceptual and comparative analysis of the different disciplinary meanings of adaptation followed by integration into a new definition oriented towards practical adaptation goals. After initial individual literature analysis, we performed collaboration-oriented activities, which included table games involving image association to stimulate creative thinking, and uncover epistemic differences, as well as co-creation of an art poster through image design and exaptation which helped us build skills in communication, co-design, and co-production. These skills were subsequently used for discussing translation, selecting suitable case studies, and collective writing. In the process, continuous feedback, constant evaluations, and reflective sessions proved pivotal. We did not explicitly strive for consensus but attempted to appreciate diverse ideas and balance them against one another and prior assumptions—a process referred to as "interdisciplinary learning" (Mansilla, 2010).

Reframing Adaptation

Adaptation is a pillar of climate action (Orlove, 2022). The concept of adaptation is used by policymakers, local organisations, and researchers in STEM and SSH, all of whom we engage with through this work. The EU's adaptation strategy refers to a set of measures developed to cope with adverse impacts of existing climate conditions and variability. The IPCC provides a widely accepted definition of adaptation, centred on broad characterisation: "In human systems, the process of adjustment to actual or expected climate and its effects, in order to moderate harm or exploit beneficial opportunities. In natural systems, the process of adjustment to actual climate and its effects; human intervention may facilitate adjustment to expected climate and its effect" (IPCC, 2023). Various types of adaptation are separately distinguished, including reactive, private and public, autonomous and planned. For each type, capacities, benefits, and costs must be robustly interpreted, compared, and operationalised. For example, technological solutions, such as air conditioning to face heat, raise concerns about energy demands and conflicts with climate mitigation goals (Mastrucci et al., 2023).

Furthermore, when it comes to climate change adaptation efforts, a closer look at the literature reveals epistemological and ethical concerns. Reasonings from SSH are often not accounted for within the STEM research on adaptation. There is also insufficient integration of diverse knowledge into policy development (Crowther et al., 2023). From a practical perspective, locally-guided and context-specific measures require more development to sufficiently address climate change exposure and vulnerability at both individual and community levels (IPCC, 2023). Despite evident progress, adaptation gaps persist by way of ephemeral benefits, fragmentation, inequality, and systemic barriers (IPCC, 2023; Orlove, 2022).

We argue that misdirected adaptation action refers to the lack of a framework which accounts for understanding and interpreting the complexity of strategies from local to global (Orlove, 2022). For this reason, we propose an integrated conceptual framework where disciplines such as Biology, including Evolutionary Biology, Social Sciences in a Sustainability context, and Philosophy of Science highlight different and complementary aspects of adaptation for practical developments. Specifically:

- Evolutionary Biology emphasises the way populations persist and change over long periods through consistent exploration of options which promote survival and compatibility with the environment (Lewontin, 1978).
- Social Sciences in a Sustainability context allow capturing human behaviour and social interactions, cultural practices, societal structures, and power dynamics and how these shape and interact with adaptation processes (Shrivastava et al., 2020).
- Philosophy of Science, with its recent notions of adaptivity and adaptation, underlines that both natural and social systems actively engage with the environment in a continuous process that not only involves stability and passive response to a perturbation but implies innovation and reorganisation (Menatti et al., 2022).

Through a process of knowledge integration, we provide a new definition of adaptation, which allows the formulation of effective policies:

Adaptation should be reframed as processes which proactively engage with disrupting climate events and involve adopting situated and relational long-term practices relating people and their ecological, social, and historical environments.

Every process of adaptation implies the improvement of a social system to its environment and should be analysed via local and global lenses (see also Menatti et al., 2022; Simonet & Duchemin, 2010).

The main policy-oriented implications of our new definition are:

- Adaptation is always context-dependent due to peoples' continuous engagement with ecological, social, and historical environments.
- Adaptation involves situated social practices and interests which policy must initially account for and subsequently evolve with.
- Adaptation pathways involve a dialogue between local and global perspectives.

We used this new framework as a lens to understand and learn from three Indian examples of adaptation practices.

Understanding Heatwave Adaptation in India

We focus our attention on India because of its success in adapting to heatwaves and because of our team's direct experiences and affiliations, which offer unique access to insights on India's heatwave solutions. Despite the enormity of the problem, with increases in frequency, duration, and intensity of heatwaves over the past decade in both areas, India's heat-related mortality rate is much lower relative to population than that of European countries (Ballester et al., 2023). This difference can be attributed to long-term evolution of heat adaptive strategies in response to extreme summers which are now rooted in India's culture and traditions. In more recent times, these include ceiling fans and desert air coolers as effective and low-energy strategies.

The EU is interested in strategies at the intersection of the social and ecological (EEA, 2013). However, our reframed concept of adaptation suggests that the social and ecological are intertwined with the historical. India provides many examples of strategies, effective and long-term oriented, which capture the alignment between the EU's interests and our interdisciplinary adaptation framework. Here we illustrate three strategies employed to adapt to urban heat, which focus on the revival of traditional climate-sensitive architecture, nature-based solutions, and women empowerment in vulnerable and deprived communities. Although these adaptation strategies have been shaped by specific climatic, ecological, and societal conditions, elements of their success may be translatable to European contexts.

Jaali Fenestrations: Revival of Traditional Climate-Sensitive Architecture

Traditional architecture from the hot regions of India provides thermal comfort by passive or natural cooling through various mechanisms, including shade and cross ventilation (Gupta, 1984). One such architectural element is *jaali* fenestrations, a latticed framework with intricate patterns of perforations which protects from the direct glare of the sun, but allows ventilation and natural light, and is aesthetically associated with Indian culture and history (Prasad et al., 2022). Recently, *jaali* are being revived in urban buildings for private dwellings and public spaces. The use of *jaali* appeals to both cultural values and climate interests, being an affordable, low-energy alternative to AC (Azmi, 2022).

Ashwath Kattes: Sacred Trees as Social Spaces and Nature-Based Solution for Heat Alleviation

Reverence and worship of trees, especially *Ficus benghalensis* and *Ficus religiosa* with their broad canopies and abundant shade, is widely prevalent across India (Nagendra & Mundoli, 2019). The *Ashwath Katte*s in Bengaluru city are enshrined spaces under an Ashwath tree (*F. religiosa*) which are protected from rapid urbanisation because they are held sacred. At *Ashwath Kattes*, a raised platform around each tree allows people to gather to meet or rest, and small vendors to sell their goods while simultaneously serving as refugia from extreme heat, especially for people in deprived communities living in cramped spaces within the city (Keswani, 2017). Studies in Bengaluru city have shown the significant potential for urban microclimate amelioration of street trees, which can reduce ambient air temperatures, road surface temperatures, and SO_2 levels (Vailshery et al., 2013), and urban green spaces, whose cooling effects can spread far beyond their boundaries (Shah et al., 2021).

Mahila Housing Trust: Women Empowerment in Deprived Communities

Mahila Housing Trust (MHT), where "Mahila" translates to "Women" in Hindi, is a successful grassroots women's organisation, established in 1994 by a union of impoverished, self-employed women workers to improve housing and infrastructure in Gujarat state (https://www.mahilahousingtrust.org/). More recently, MHT has been collaborating with engineers, architects, and policymakers as it works proactively towards alleviating the impact of rising temperatures on the health of disadvantaged groups. Under their Climate Resilience Programme, MHT empowers women by training them to become *Climate Saathis* (partners), who visit various communities and households to raise residents' awareness and bolster resilience by advocating for climate-friendly solutions such as energy-efficient household appliances and facilitating the adoption of solar reflective paints for cooler roofs. The MHT programme, which exemplifies the feminist movement to face climate change (Turquet et al., 2023) received the prestigious 2019 United Nations Global Climate Action Awards at COP25, and today continues its work across India, Bangladesh, and Nepal.

The Indian mix of traditional and contemporary adaptation strategies for heatwaves offers several learning possibilities (Fig. 4.1), discovered through the intersection of disciplines. The successes achieved through empowering communities, especially women, and increasing residents' awareness of climate change are understood through the joint lens of Social Ecology with Philosophy of Science, which focuses on knowledge, perspective, and justice. Similarly, through pairing the disciples of Biology and History, green spaces and traditional climate-sensitive architecture are revealed as valued sources of affordable, low-energy solutions for disadvantaged communities. These examples show how the combination of STEM and SSH within an interdisciplinary analysis can provide innovative knowledge for guiding adaptation actions. In the next section, we consider the theoretical and practical steps to apply this knowledge to the European context.

Translating Adaptation for Europe

Global sharing of adaptation strategies may be pivotal for successful climate action. Existing experience from different regions can fill significant gaps in knowledge and understanding and drive innovation. However, knowledge transfer from one situation (study case) to another (target case) faces considerable challenges, making it a highly debated issue within sustainability sciences (e.g., Adler et al., 2018; Cartwright, 2012). Assessment of the potential for transfer between cases is often limited by unseen differences and misalignment of researcher and policymaker interests. In addition, premature application of study results and insufficient consideration of local needs, capacities, and values can result in adaptation action failure or even maladaptation.

For this reason, along with our reconceptualisation of adaptation, we see knowledge transfer as only part of a more complex process, *"Translation"*. In linguistics, translation consists of the interpretation and articulation of meaning across language contexts. Similarly, for climate action research, rather than mere transfer, we see translation as a process that involves the extraction of knowledge from a source context followed by application within a target context. We can ask, how can India's successful climate responses be *translated* into successful climate action

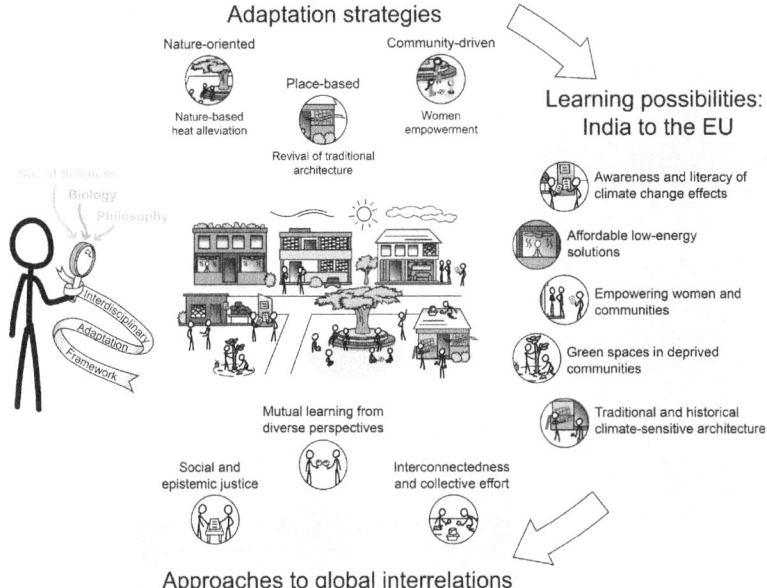

Fig. 4.1 Interdisciplinary adaptation framework as a magnifying lens which we used to understand adaptation to climate change in India. Following examination of multiple "Adaptation strategies" for extreme temperatures, a series of "Learning possibilities" for the EU emerged which highlight crucial "Approaches to global interrelations". Graphic by Corey Bunce

in the EU and vice versa? We propose that translation involves an interdisciplinary examination of both situations so that common ground can be found and solutions shared intergeographically and interculturally through reimagining practices and tailoring policy according to local options and resources.

As a team fulfilling these guidelines, we re-examined our examples of heatwave adaptation to assess the potential for translation between India and similar situations in the EU. Based on our understanding of the ways each case constitutes successful adaptation within India, we identified policy opportunities for EU urban areas within and across individual, community, and national scales:

- Empowering communities, specifically women, to build infrastructure and resources for transformative change leads to multi-level capacities for climate change awareness and adaptation action.
- Increasing green spaces and native nature-based solutions in disadvantaged communities is a way to combat heat while promoting social cohesion and cultural engagement.
- Ensuring access to affordable, climate-sensitive solutions which appeal to traditional and historical architecture allows communities to find strategies which fit socio-cultural, environmental, and economic specificities.

In line with our adaptation framework, these opportunities involve relational and context-sensitive structures. For translation, the Climate-ADAPT website tool[1] acknowledges that "adaptation must be tailored to the scale required by the climate change challenge (e.g., national/regional/local/sectoral/cross-border) and solutions need to be modified for individual situations, also addressing responsibilities and financing". However, there are scant resources to help researchers perform the task of translation.

Our analysis is a first attempt which shows that to face heatwaves, the EU should analyse contexts from an interdisciplinary perspective, relating Social Sciences, Philosophy of Science, and STEM fields. Consequently, the resulting policies will prioritise socio-spatial vulnerabilities whose understanding enables contextual, affordable, awareness-driven, and nature-based interventions. Epistemic perspectives from the Global South can be a source of learning, inspiration, and diverse knowledge. We emphasise in this sense the opportunity for mutual learning and justice between the Global North and Global South.

Conclusions and Recommendations

Increasing temperatures and heatwaves in Europe pose imminent threats to ecosystems, infrastructures, and health. In this chapter, we have argued that accomplishing the EU's goals for adaptation to climate change, and meeting the challenges raised by the EUCRA, requires a reframing of the concept of "adaptation", in particular to address risks faced by vulnerable populations. Through an interdisciplinary process, we developed an optimal reframing which understands adaptation as consisting of situated and relational long-term processes involving people and their ecological, social, and historical environments.

To demonstrate the value of our interdisciplinary adaptation framework, we focused our attention on adaptation strategies in India, drawing on the similarity between the current heat crisis in Southern Europe and India's history of extreme summers. We showed how practices within India, including historical architecture and green spaces as well as efforts in women empowerment, can be understood as learning possibilities for Europe. This analysis shows how the EU can benefit from drawing upon the Global South for its adaptation policies.

In order to develop European policies from adaptation successes in the Global South, we outlined a framework for "translation" which recognises the relational and context-sensitive aspects of case knowledge. We identified policy opportunities which appear at the intersection of epistemic perspectives and scales, highlighting the importance of interdisciplinarity in the construction of adaptation resources and tools. Moreover, our results demonstrate the potential for the EU to overcome challenges of translation by engaging in global collaboration through frameworks of interconnectedness, mutual learning, and social and epistemic justice (Fig. 4.1).

Acknowledgements We would like to thank Dr Guido Caniglia and the Konrad Lorenz Institute (KLI) in Klosterneuburg for the opportunity to be part of the diverse, vibrant, and supportive community. We are extremely grateful to Guido for his contribution as facilitator in this project. He has been a continuous source of inspiration and insights for our theoretical, philosophical, and scientific reasoning. We, as a research team, have learned a lot from him about how to work effectively in an interdisciplinary and proactive fashion.

Notes

1. https://climate-adapt.eea.europa.eu/en/knowledge/tools/adaptation-support-tool/step-0-3

REFERENCES

Adler, C., Hirsch Hadorn, G., Breu, T., Wiesmann, U., & Pohl, C. (2018). Conceptualizing the transfer of knowledge across cases in transdisciplinary research. *Sustainability Science, 13*(1), 179–190.

Azmi, F. T. (2022, September 21). *How India's lattice buildings cool without air con*. BBC Future. https://www.bbc.com/future/article/20220920-how-indias-lattice-buildings-cool-without-air-con

Ballester, J., Quijal-Zamorano, M., Méndez Turrubiates, R. F., Pegenaute, ... & Achebak, H. (2023). Heat-related mortality in Europe during the summer of 2022. *Nature Medicine, 29*(7), 1857–1866.

Cartwright, N. (2012). Presidential address: Will this policy work for you? Predicting effectiveness better: How philosophy helps. *Philosophy of Science, 79*(5), 973–989.

Crowther, A., Foulds, C., & Robison, R. (2023). *A review of the Climate-Energy-Mobility landscape through 10 Social Sciences and Humanities literature briefs*. SSH Centre.

European Environment Agency (EEA). (2013). Adaptation in Europe—Addressing risks and opportunities from climate change in the context of socio-economic developments (Report No 3/2013).

European Environment Agency (EEA). (2024). The EUCRA European Climate Risk Assessment (Report No. 1/2024).

Gupta, V. (1984). Indigenous architecture and natural cooling. *Energy and Habitat*, 41–58.

IPCC. (2023). *Climate change 2023: Synthesis report. Contribution of working groups I, II and III to the sixth assessment report of the intergovernmental panel on climate change* [Core Writing Team, H. Lee and J. Romero (eds.)]. IPCC, Geneva, Switzerland, pp. 35–115.

Keswani, K. (2017). The practice of tree worship and the territorial production of urban space in the Indian neighbourhood. *Journal of Urban Design, 22*(3), 370–387.

Lewontin, R. (1978). Adaptation. *Scientific American, 239*(9), 212–230.

Mansilla, V. B. (2010). Learning to synthesize: The development of interdisciplinary understanding. In R. Frodeman, J. T. Klein, & C. Mitcham (Eds.), *The Oxford Handbook of Interdisciplinarity* (pp. 288–306). Oxford University Press.

Mastrucci, A., Niamir, L., Boza-Kiss, B., Bento, N., & van Ruijven, B. (2023). Modeling low energy demand futures for buildings: Current state and research needs. *Annual Review of Environment and Resources, 48*, 761–792.

Menatti, L., Bich, L., & Saborido, C. (2022). Health and environment from adaptation to adaptivity: A situated relational account. *History and Philosophy of the Life Sciences, 44*(3), 1–28.

Nagendra, H. (2018). The global south is rich in sustainability lessons. *Nature, 557*, 485–488.

Nagendra, H., & Mundoli, S. (2019). *Cities and canopies: Trees in Indian cities.* Penguin Random House India Private Limited.

Orlove, B. (2022). The concept of adaptation. *Annual Review of Environment and Resources, 47*(1), 535–581.

Prasad, R., Tandon, R., Verma, A., Sharma, M., & Ajmera, N. (2022). Jaali a tool of sustainable architectural practice: Understanding the feasibility and usage. *Materials Today: Proceedings, 60*, 513–525.

Shah, A., Garg, A., & Mishra, V. (2021). Quantifying the local cooling effects of urban green spaces: Evidence from Bengaluru. *India. Landscape and Urban Planning, 209*, 104043.

Shrivastava, P., Smith, M. S., O'Brien, K., & Zsolnai, L. (2020). Transforming sustainability science to generate positive social and environmental change globally. *One Earth, 2*(4), 329–340.

Simonet, G., & Duchemin, E. (2010). The concept of adaptation: Interdisciplinary scope and involvement in climate change. *Surveys and Perspectives Integrating Environment and Society, 3*(1), 392–401.

Turquet, L., Tabbush, C., Staab, S., Williams, L., & Howell, B. (2023). *Feminist climate justice: A framework for action. Conceptual framework prepared for Progress of the World's Women series.* UN-Women.

Vailshery, L. S., Jaganmohan, M., & Nagendra, H. (2013). Effect of street trees on microclimate and air pollution in a tropical city. *Urban Forestry & Urban Greening, 12*(3), 408–415.

Open Access This chapter is licensed under the terms of the Creative Commons Attribution 4.0 International License (http://creativecommons.org/licenses/by/4.0/), which permits use, sharing, adaptation, distribution and reproduction in any medium or format, as long as you give appropriate credit to the original author(s) and the source, provide a link to the Creative Commons license and indicate if changes were made.

The images or other third party material in this chapter are included in the chapter's Creative Commons license, unless indicated otherwise in a credit line to the material. If material is not included in the chapter's Creative Commons license and your intended use is not permitted by statutory regulation or exceeds the permitted use, you will need to obtain permission directly from the copyright holder.

CHAPTER 5

Advancing Epistemic Justice with Local Knowledge: A Process Indicator for EU Climate Adaptation Policymaking

Hernán Bobadilla, Giuseppe Di Capua, Chris Hesselbein, Silvia Peppoloni, and Federico Lampis

Policy Highlights To achieve the recommendation stated in the title, we propose the following:

- EU climate adaptation policies need to further integrate local knowledge to advance epistemic justice and ensure their success.
- A process indicator is proposed to advance epistemic justice along three main dimensions, namely distributive, participatory, and recognitional epistemic justice.

The information and views set out in this chapter are those of the author and do not necessarily reflect the official opinion of the European Commission.

H. Bobadilla (✉) · C. Hesselbein
Department of Mathematics, Politecnico Di Milano, Milan, Italy
e-mail: hernanfelipe.bobadilla@polimi.it

C. Hesselbein
e-mail: christopher.hesselbein@polimi.it

- The indicator serves to assess and evaluate critical ex-ante (problem framing) and ex-post (appraisal of the policy's initial design) aspects of epistemic justice in policymaking.
- The implementation of the indicator will enhance political accountability, fill existing gaps in scientific knowledge at smaller spatial scales, and foster trust among stakeholders.
- The inclusion of multiple types of knowledges and disciplines in policymaking leads to more effective and just climate policies.

Keywords Epistemic Justice · Local Knowledge · Climate Adaptation · Process Indicator · EU policymaking

Introduction

Adaptation is of paramount importance in dealing with the wide-ranging effects of climate change at the local, national, and global levels. This involves processes of adjustment to current and future climates to reduce exposure and vulnerability. To be successful, adaptation relies on two considerations. First, adaptation requires state-of-the-art, evidence-based knowledge about climate and social-ecological systems to ensure its efficiency and feasibility. Second, adaptation calls for fair processes of planning and policymaking to ensure justice. Both considerations—knowledge and justice—are combined in an increasingly recognised political aim,

---

G. Di Capua · S. Peppoloni
Istituto Nazionale di Geofisica e Vulcanologia, Rome, Italy
e-mail: giuseppe.dicapua@ingv.it

S. Peppoloni
e-mail: silvia.peppoloni@ingv.it

C. Hesselbein
Department of Management Engineering, Politecnico di Milano, Milano, Italy

F. Lampis
Directorate-General for Research and Innovation, European Commission, Brussels, Belgium
e-mail: fedelampis@gmail.com

namely "epistemic justice", which encompasses criteria and standards that seek to ensure fair and equal recognition, representation, and participation by diverse actors in processes of knowledge production.

The main aim of this chapter is to offer practical recommendations to policymakers while simultaneously underscoring the crucial role of epistemic justice in climate change adaptation and emphasising the importance of locally sourced knowledge. We therefore propose the introduction of a process indicator for the evaluation of epistemic justice, specifically for the degree to which locally available knowledges and practices are acknowledged, supported, reinforced, as well as integrated into local and national policies (and potentially beyond). The inclusion of local knowledge is crucial for: (i) filling existing gaps in scientific knowledge at small spatial scales, (ii) mitigating potential systemic biases that are inherent to the scientific approach (e.g., ontological assumptions, institutionalised cultural norms, and validation standards), (iii) fostering trust between local stakeholders, scientists, and policymakers, (iv) ensuring that adaptation policies consider a diversity of local perspectives and needs, and (v) making adaptation policies more actionable and effective. We suggest implementing the indicator by integrating it into the European Commission's Better Regulation (BR) framework, which seeks to ensure that legislation is evidence-based, simpler, better, and inclusive of all relevant stakeholders affected by ensuing policies.

We substantiate our policy recommendation in the context of a leading European initiative for climate adaptation, namely the EU Strategy on Adaptation to Climate Change. The EU Adaptation Strategy has four main objectives: to make adaptation smarter, faster, more systemic, and to step up international actions for climate resilience. All EU member states are obliged to prepare and implement national energy and climate plans by 2024 in line with the EU-wide strategy to become climate-neutral and resilient by 2050. Despite its merits, we identify two significant problems with the EU Adaptation Strategy that threaten its potential success.

First, the goal of faster adaptation conflicts with the necessity of investing the required time to engage in the laborious tasks that foster justice in systemic changes. Typically, systemic changes call for the development and implementation of adaptation plans and actions at all levels of governance. This demands great concerted efforts and resources, especially in terms of time, which allow for deliberative, noncoercive processes of discussion and negotiation among various stakeholders. Moreover, an explicit priority of the EU Adaptation Strategy for promoting systemic change is the development of local adaptation actions. To formulate and implement policies that foster just local adaptation actions, it is essential to fully engage with the specificities of each locality, considering both its internal commonalities and heterogeneities, as well as the potential conflicts and synergies with other localities. What is more, some localities might require faster adaptations due to more severe climate risks whereas other areas might have more margin for slower adaptations, thus further highlighting the necessity of developing granular approaches.

Second, although all EU member states are obliged to prepare and implement national energy and climate plans, the efficient implementation of national strategies and the accomplishment of climate resilience largely rely on the actionability of these plans (i.e., how meaningful and compelling they are). Both the integration of local knowledge and the achievement of epistemic justice radically improve the prospects of adaptation policies in terms of their actionability and subsequent likelihood of success. For instance, the integration of local knowledge promotes a positive affective response (shaped by past experiences), which simultaneously fosters the public's adoption of policies as well as their legitimisation.

This chapter brings together experts on Philosophy and Sociology of Science, Political Science, Geological Hazards, and Geoethics. Through an iterative process of remote and in-person meetings over the course of six months, we have sought to combine our respective disciplinary backgrounds and expertise both in terms of fieldwork experiences and theoretical proficiencies. During our collaboration, we shared and discussed policy documents and the available literature from our own and adjacent academic fields, presented and debated relevant approaches and frameworks, and set up a working document for drafting overviews of core topics and issues, which were subsequently commented on and redrafted as necessary. Based on this process, we establish an interdisciplinary consensus to substantiate our claim that epistemic justice and local knowledge are mutually dependent factors that underpin fair, actionable, and efficient climate adaptation policies. Two disparate, yet related, bodies of literature guide our policy recommendation, namely (i) transdisciplinary research on local, traditional, and indigenous knowledge, and (ii) philosophical research on epistemic justice. In addition to this theoretical knowledge, we draw on our combined experiences in the field in terms of (i) direct engagement with local communities, especially in terms of communication and management of geological hazards, (ii) the development of practical geoethical principles to guide interactions with local communities, and (iii) involvement in policymaking processes at the EU level.

In what follows, we first explain the concept of epistemic justice and how it might be instrumentalised for policymaking. Next, we briefly elaborate on the importance of local knowledge. Last, we justify the use of our process indicator and discuss the specific problems it seeks to address. To access our indicator prototype, see the Appendix.

Why Is Epistemic Justice Pertinent?

The term "epistemic *in*justice" emerged in the early 2000s from the work of feminist philosophers working at the interface of ethics and epistemology (e.g., Fricker, 2007). However, its origins can be traced to long-standing problems in political philosophy and ethics. A focus on *"in*justice" (as opposed to justice) is not arbitrary: "Injustices" are what individuals and groups experience in their daily lives, whereas "justice" is a theoretical ideal that is frequently contested and elusive to achieve. Epistemic injustices can be characterised as wrongs to individuals and groups in their capacity as holders and seekers of knowledge. These wrongs include (but are not limited to) the undervaluing, silencing, and exclusion of various knowledges. We conceive epistemic justice as the progressive reduction of these wrongs in multiple ways according to contextually dependent values and norms. In this sense, we claim that achieving epistemic justice is an incremental and relational process that is open to renegotiation and adjustment through public deliberation (Sen, 2009).

To instrumentalise the notion of epistemic justice, we distinguish three mutually supporting components (see Fig. 5.1). First, epistemic justice can be understood in *distributive* terms: who gets what and how. Our indicator asks questions concerning the distribution of various aspects of knowledge production processes, such as services, information, skills, and infrastructure, among others. Our scoring system is "prioritarian", which means that our indicator values distributions that benefit those most affected by epistemic discrimination and marginalisation. We made this decision in order to highlight injustices towards local communities because their knowledges have historically been neglected or undervalued.

Second, epistemic justice can be understood in *participatory* terms. Participation means that members of society have the opportunity to

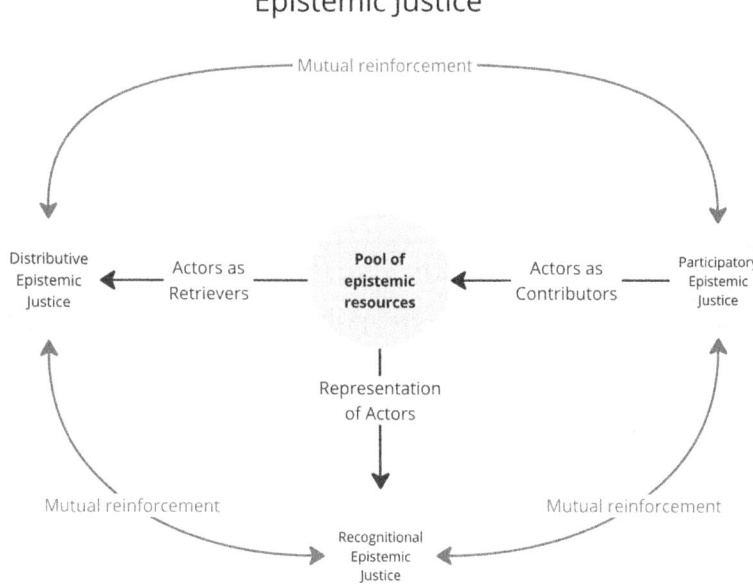

Fig. 5.1 Epistemic justice as composed of distributive, participatory, and recognitional justice (cf. Mathiesen, 2015)

communicate their views and experiences in processes of shared decision-making. Participatory epistemic justice could be advanced in a "bottom-up" fashion, i.e., through the correction of prejudices, discrimination, and abuses in relations among individuals. However, given our focus on policymaking, we opt for a "top-down", institutional approach. Our indicator seeks to address various forms of participatory epistemic injustice, but two are worth highlighting. First, testimonial injustices, in which local communities receive less credibility than they deserve because of systemic prejudice in institutional contexts. Second, hermeneutical injustices, in which local communities are unable to render their experiences and perspectives intelligible, either to themselves or to others, because systemic discrimination has prevented them from establishing or even finding suitable means (adapted from Fricker, 2007).

Third, epistemic justice can be understood in *recognitional* terms. This means ensuring the fair and accurate representation of all members of society in the broader pool of knowledge. Accurate representation may be facilitated by fairer participation but does not necessarily ensure it: Individuals and groups may participate epistemically while withholding certain distinctive experiences out of historical humiliation, disrespect, lack of social esteem, cultural dominance, and status hierarchy (Honneth, 2004). Local knowledge is particularly susceptible to being ignored. Small and local communities, with their own distinctive experiences and knowledges, frequently face resistance (if not active silencing and abuse) from dominant groups in the shared pool of knowledge (Naess, 2013). Our indicator considers the degree of representation by local communities in order to strengthen their level of epistemic recognition in processes involving different stakeholders, relations, conflicts, and uncertainties.

Why Is Local Knowledge Essential for Climate Adaptation?

The literature on local, traditional, and indigenous knowledges and "ways of knowing" is informed by multiple disciplines and diverse approaches from the SSH. Moreover, there are many peoples and communities across the world with varied cultures and understandings of their specific ecological contexts and historical pathways that are profoundly different. It is therefore impossible to provide a single and unified definition of all these different ways of knowing one's local environment and community. One can nevertheless identify some common characteristics of local knowledges: they emerge from close interaction and association with the land and its associated social-ecological systems; they are cumulative and collectively developed and (continue to be) transmitted across generations; and they represent a cohesive bundle of culturally specific practices, values, beliefs, and worldviews about the relationship between humans and their environment (Agrawal, 1995; Naess, 2013).

There are important differences between the qualifiers "local", "traditional", and "indigenous" that cannot be discussed here. We chose the term "local" because it appears to be the most fitting for our needs in the European context, where indigenous groups are relatively sparse, though with some important exceptions. Rather than defining the term "local" in relation to geographical distances or existing geopolitical boundaries, we chose to let the term be defined by the shared level of exposure and vulnerability of those affected by events related to climate change in a certain geographical area. In the European context, the term "local" can thus apply to particular communities of practice that value and share common concerns, such as, for example, inhabitants of flooded lowlands on the coast of the North Sea or fisherpeople around the Adriatic Sea, whose livelihoods and cultural/natural heritage are exposed to common threats.

Local knowledge is possessed by a wide variety of groups, whether professionals, such as farmers and forest caretakers, or lay/amateur groups and citizens whose activities and local presence endow them with an awareness of changes affecting local social-ecological systems. It is crucial to note that such groups reflect a heterogeneity of knowledge that emerges from their specific engagement with the environment as well as their respective interests, values, and identities. This is the reason why all these local groups will have different epistemic positions in relation to climate risks. The inclusion of local knowledge in climate change adaptation strategies has been deemed an essential means for filling gaps in scientific knowledge at smaller spatial scales, mitigating systemic biases that are inherent to the scientific approach, fostering trust between stakeholders, scientists, and policymakers, and ensuring that adaptation policies address local issues or wider societal concerns (Jasanoff, 2021; Kieslinger et al., 2019; Klenk et al., 2017; Naess, 2013; Wheeler & Root-Bernstein, 2020).

It must be noted, however, that our goal of fostering the integration of local knowledge into institutional frameworks and processes, particularly of large political entities such as the EU, bears certain risks that need to be addressed. Institutionalising local knowledge, for example, could lead to a more hermetic and fixed conceptualisation of local knowledge that does not correspond to its variability or flexibility on the ground, and moreover, cause local knowledge to become subsumed or overshadowed by larger imperatives and therefore undo meaningful integration. Our indicator is not immune to these risks: as with any tool, its merits can be undermined by incomplete implementation.

Why Use a Process Indicator for Achieving Epistemic Justice in EU Policies?

Indicators are tools that simplify the description of complex phenomena into a few dimensions for qualitative/quantitative and standardised analysis, serving as support for designing or amending legal frameworks. We acknowledge that any indicator of epistemic justice will inevitably fail to cover all existing unjust relations in any local domain. A top-down indicator of epistemic justice for policymaking does not replace the bottom-up relations of care and trust that enable justice to emerge organically. Nevertheless, it can help lessen unequal power dynamics by balancing scientific expertise with the involvement of local knowledge. Climate adaptation calls for action on multiple fronts. Our proposed indicator is only one of many ways to advance epistemic justice in climate adaptation policies, one that is targeted directly at policymakers.

Having stated this, we argue that our indicator possesses evident strengths because it is: (i) generalisable; (ii) concrete and actionable; and (iii) a measurement tool that simultaneously provides guidance on how to increase epistemic justice in EU policymaking. EU climate policymaking often focuses on setting quantifiable targets for drivers of climate change such as level of emissions or energy use. While certain social goals (including attention to epistemic justice) may be considered horizontally during the policymaking process, assessing a policy's success would usually focus on measuring these quantifiable targets. Our indicator seeks to address this gap by providing policymakers with a tool to systematically measure the level of epistemic justice in EU climate adaptation policies. The tool is designed as a "process" (as opposed to "outcome") indicator in the form of a checklist to be used by policymakers. The indicator is not case-dependent, which makes it applicable across multiple policies in the climate adaptation realm. Its use can provide policymakers with instant feedback on the successful integration of epistemic justice considerations within the initiative that is being developed, guiding the drafting of more just policies. Lastly, its structure and functioning allow it to be integrated into the European Commission's Better Regulation Toolbox and thus feed into the EU's existing policymaking workflow, making the indicator concrete and actionable.

In developing the indicator, we conceptualise the policymaking process as consisting of two main stages. First, an *ex-ante* stage (i.e., before the policy is designed) in which the problem is explored and framed with relevant stakeholders. Second, an *ex-post* stage (i.e., after the policy is designed but not yet finalised and implemented) in which the prospective policy is tested in consultation with relevant stakeholders. Our indicator assesses the three above-mentioned components of epistemic justice (distributive, participatory, and recognitional) during these two stages of policymaking. The assessment is performed using a checklist with binary and multiple-choice questions. The answers are then counted and evaluated, enabling a score by module and stage. As a tool for advancing epistemic justice in policymaking, our indicator is directly intended for policymakers. However, the indicator could in principle also be used by other stakeholders to hold policymakers accountable or to challenge policies. For example, stakeholders might critically assess or even denounce aspects of the policymaking process using the standardised metric provided by the indicator. Ultimately, successful implementation of the indicator will depend on various contextual aspects, including the infrastructural capacities and limitations of territorial and local institutions. Our prototype, together with further instructions, can be found in the Appendix.

Conclusions and Recommendations

Our process indicator allows for the advancement of epistemic justice, by providing criteria for assessing the extent to which local knowledge informs policymaking, both in problem framing and policy appraisal stages. We assert—based on our interdisciplinary collaboration—that this makes climate adaptation policymaking not only more just, but also ensures that policies are more actionable and efficient in addition to making policymakers more accountable.

Our process indicator contributes directly to addressing specific gaps in the EU Better Regulation (BR) framework. In principle, the BR framework already contains the seeds for advancing epistemic justice because it states that all interested parties should be able to participate in policymaking. Our indicator addresses the shortcomings of the BR framework, specifically its failure to provide concrete and comprehensive instructions for advancing epistemic justice at the ex-post stage as well as to provide any ex-ante evaluation mechanism. The integration of our indicator into the BR framework would strengthen the mechanisms for collecting evidence from diverse stakeholders and evaluating their implementation. We therefore suggest that our process indicator is added to the BR toolbox to explicitly guide policymaking and facilitate the development of more epistemically just policies. Subsequently, the scope of epistemic justice can be expanded to policymaking in areas beyond climate adaptation.

The next steps would include encouraging territorial and local institutions to exploit this opportunity at the local level. For example, we hope that programmes such as the Regional Hubs Network will make use of our indicator to assess how epistemically just their processes are and to make the proper amendments if necessary. Only once local knowledges and concerns are better integrated into broader policy frameworks can more effective and just climate policies be properly enacted.

Acknowledgements Project 101105236-UN3 (HORIZON-MSCA-2022-PF): Understanding Under Uncertainty.

Appendix

Bobadilla, H., Di Capua, G., Hesselbein, C., Peppoloni, S. & Lampis, F. (2024). Epistemic justice indicator: An annotated prototype. *Zenodo.* https://doi.org/10.5281/zenodo.13712721

For an updated version of the indicator, please visit this live document: https://docs.google.com/document/d/1MehlBjdoLmr5QYts8AfEsdJMmG_poVCo/edit

References

Agrawal, A. (1995). Dismantling the divide between indigenous and scientific knowledge. *Development and Change, 26*(3), 413–439.

Fricker, M. (2007). *Epistemic injustice: Power and the ethics of knowing*. Oxford University Press.

Honneth, A. (2004). Recognition and justice: Outline of a plural theory of justice. *Acta Sociologica, 47*(4), 351–364.

Jasanoff, S. (2021). Knowledge for a just climate. *Climatic Change, 169*(36), 1–8.

Kieslinger, J., Pohle, P., Buitrón, V., & Peters, T. (2019). Encounters between experiences and measurements: The role of local knowledge in climate change research. *Mountain Research and Development, 39*(2), 55–68.

Klenk, N., Fiume, A., Meehan, K., & Gibbes, C. (2017). Local knowledge in climate adaptation research: Moving knowledge frameworks from extraction to co-production. *Wiley Interdisciplinary Reviews: Climate Change, 8*(5), e475.

Mathiesen, K. (2015). Informational justice: A conceptual framework for social justice in library and information services. *Library Trends, 64*(2), 198–225.

Naess, L. O. (2013). The role of local knowledge in adaptation to climate change. *Wiley Interdisciplinary Reviews: Climate Change, 4*(2), 99–106.

Sen, A. (2009). *The Idea of Justice*. Harvard University Press.

Wheeler, H. C., & Root-Bernstein, M. (2020). Informing decision-making with Indigenous and local knowledge and science. *Journal of Applied Ecology, 57*(9), 1634–1643.

Open Access This chapter is licensed under the terms of the Creative Commons Attribution 4.0 International License (http://creativecommons.org/licenses/by/4.0/), which permits use, sharing, adaptation, distribution and reproduction in any medium or format, as long as you give appropriate credit to the original author(s) and the source, provide a link to the Creative Commons license and indicate if changes were made.

The images or other third party material in this chapter are included in the chapter's Creative Commons license, unless indicated otherwise in a credit line to the material. If material is not included in the chapter's Creative Commons license and your intended use is not permitted by statutory regulation or exceeds the permitted use, you will need to obtain permission directly from the copyright holder.

CHAPTER 6

Linking Vulnerability to Heatwaves and Public Health: Indicators for EU Policies on Energy Renovation of Residential Buildings

Ángela Lara-García, Carlos Rivera-Gómez, Claudia Núñez-Rivera, Carmen Galán-Marín, and Estrella Candelaria Cruz-Mazo

Policy Highlights To achieve the recommendation stated in the title, we propose the following:

- Recognising heatwave risks as a public health problem, energy renovation of buildings must focus on the most vulnerable groups.
- Renovation programmes require an integrative approach, interdisciplinary teams, and socio-educational initiatives such as "energy coaching".

Á. Lara-García (✉) · C. Rivera-Gómez · C. Núñez-Rivera
Territorial Structures and Systems Research Group, University of Seville, Seville, Spain
e-mail: anglargar@us.es

C. Rivera-Gómez
e-mail: crivera@us.es

- A multi-criteria framework is proposed for the assessment of heatwaves vulnerability integrating biophysical and socio-economic factors.
- Main factors of vulnerability are identified: income level; population over 65 (mostly elderly women living alone); educational level; ageing and quality of buildings; and urban greening.
- Quality and availability of data on health, building quality, and energy use are essential to effective prioritisation of funding for renovation.

Keywords Vulnerability · Heatwaves · Energy renovation buildings · Adaptation · Public health

Introduction

As a consequence of the high concentration of population, services, and infrastructure, the impacts of climate change will be felt harshly in the urban scenario. The Urban Heat Island (UHI) effect exacerbates climate aftermaths. This brings significant technical challenges to ensuring the thermal comfort of an aged residential stock. As climate change becomes a determining factor in urban planning, new policies are being developed to anticipate its effects. While previous EU policies for buildings focused on energy efficiency, *"the need for climate change adaptation in buildings is increasingly reflected in the EU policy landscape"* (EC, 2023a,

---

C. Núñez-Rivera
e-mail: cnrivera@us.es

C. Rivera-Gómez · C. Galán-Marín
Faculty of Architecture, University of Seville, Seville, Spain
e-mail: cgalan@us.es

C. Núñez-Rivera · E. C. Cruz-Mazo
Department of Physical Geography and Regional Geographic Analysis, Geography and History Faculty, University of Seville, Seville, Spain
e-mail: ecruz@us.es

3). This need has been singled out in the European Green Deal, specifically in the EU's Renovation Wave Programme. As a mitigation strategy, energy efficiency of buildings prioritises those contexts with the highest energy consumption, often those under cold conditions. However, when building renovation is considered an adaptation strategy, the focus must shift to the contexts where the conditions of vulnerability are most significant. Despite the EU having relatively high adaptive capacity to heat risks, there are significant differences across countries, with the Mediterranean sub-region being particularly vulnerable.

Adaptation is linked to the concept of climate vulnerability—the predisposition to be adversely affected by changes in climate conditions (IPCC, 2022). Vulnerability involves a variety of biophysical and socio-economic factors related to exposure, sensitivity, or susceptibility to impacts, as well as coping and adaptive capacity. Several works have investigated the factors determining urban vulnerability in the Southern EU (Hernández et al., 2018), including vulnerability to heat (VH) (Domene et al., 2022; López-Bueno et al., 2020).

The group of vulnerable people to extreme temperature or heatwave conditions usually includes elderly or young citizens, who suffer from certain physical or mental ailments, or those in precarious socio-economic conditions (De la Osa, 2016). In fact, Eurostat (2024) reports that the elderly population (over 65 years) constitutes 21.3% of the EU's total population and is projected to rise to 32,5 % by 2100. Heat accounted for over 80% of the EU's average mortality rate from climate-related events between 2010 and 2020 (IPCC, 2022). Furthermore, the energy poverty associated with worsening climatic conditions in summer already affects 30% of the population in the Mediterranean region, with a significant increase expected by mid-century (Bouzarovski, 2014).

In this chapter, we scrutinise VH in the Mediterranean city of Seville and how it relates to the prioritisation of funds for energy renovation of buildings (ERB) from a climate adaptation perspective. Our analysis exposes several challenges that demand future attention in the EU's Renovation Wave Programme, specifically in the focus area of "Tackling energy poverty and worst-performing buildings" (EC, 2024). We devise potential implementation strategies for addressing these challenges within specific policies and actions, both at the European and local level, with a special focus on public health.

Methodology

Our cross-disciplinary process engaged architects with long-term expertise in the assessment of thermal comfort and geographers focused on climatic risks and housing policies. We applied a mixed-methods approach comprising:

- Literature review on VH factors and analysis of ERB policies.
- An expert seminar was held on January 12, 2024 at the University of Seville entitled "Vulnerability to heatwaves and energy renovation in Seville". The goal was to promote a multidisciplinary dialogue between researchers and professionals from different backgrounds (geographers, environmentalists, architects, health, etc.). Additionally, social and institutional actors directly involved in local and regional policies on residential vulnerability participated in a discussion panel. See the programme in Appendix.
- A multi-criteria framework integrating social, environmental, and building factors for the assessment of VH at a sub-city district scale. The choice of statistical indicators is based on a literature review as described in Appendix. The indicator set is tested in the case study of Seville, a city with 6 of the 15 poorest urban districts in Spain, where 31.8% of heat-related deaths are caused by climate change.

Challenges and Recommendations for ERB Policies from an Adaptive Approach to Climate Change

Through the multidisciplinary dialogue group during the expert seminar, we were able to examine diverse viewpoints on the impact of escalating heatwaves. In this section, we outline the main reflections we draw from the seminar in the form of key challenges for EU policies.

Climate Change as a Public Health Crisis

One of the most intricate challenges we face in relation to climate change is its recognition as a public health problem (WHO, 2023). This perspective would help prioritise political actions to adapt communities to changes that are already taking place. Historically, warm seasons in the EU had limited health risks. However, this changed with the onset of severe heatwaves at the beginning of the century that now recur annually. In addition, heat protection strategies are more complex than those against cold. It requires passive strategies in buildings to counteract the effects of indoor overheating that are difficult to implement. For this reason, it is imperative to focus on reducing city temperatures and neutralise the added effects of the UHI phenomenon.

During the last decades, the unprecedented intensity of summer heat has significantly increased excess mortality (Ballester et al., 2023). This raises the need for refining forecasting criteria for high-temperature episodes and for redefining the concept of heatwaves itself. Our interdisciplinary dialogue highlighted the importance of consistently redefining heatwaves from both a climatic and health perspective, unifying alert criteria. Moreover, this redefinition should prioritise and assist the most vulnerable populations, establishing tailored emergency protocols and preventive measures.

We define VH as the predisposition of a community to be adversely affected by an extreme heat event, considering its susceptibility and exposure to this event, and the capacity to absorb the effects and adapt to reduce them in the future. Strengthening the management of heatwave risks requires integrating both social and contextual determinants of VH (IPCC, 2022) linked to citizens' health. For this purpose, it is important to consider and weigh the influence of several aspects:

- Social, educational, and economic VH factors versus health problems linked to climate change.
- Difference between deaths due to heat (direct) and deaths due to heat effects (indirect, increased mortality, but linked to previous pathologies), considering that heat stroke represents only 2–3% of mortality and morbidity associated with high temperatures.
- Inequities as key elements in the framework of social determinants of health (Barton & Grant, 2006), aligned with the definition of VH.
- Importance of both vulnerability and exposure factors in health risks. Personal factors include age, diseases, treatments, daily habits, jobs, and unemployment. Environmental factors due to socio-economic conditions include poorly insulated housing, difficulties in cooling, urbanised environment, and urban green infrastructure. Additional factors include local climate, early warning systems, and health care services (De la Osa, 2016).

- Specific protocols in primary health care services for the detection of people who are vulnerable to heatwaves.

Effective Policies Against Energy Poverty

Living in inefficient buildings is often correlated with energy poverty and social problems (EC, 2020). In this regard, a particularly sensitive challenge is how to define the limits and conditions of energy poverty (Croon et al., 2023). As the IPCC's Sixth Assessment Report (IPCC, 2022, 26) recognises: "*Inequity and poverty also constrain adaptation, leading to soft limits and resulting in disproportionate exposure and impacts for most vulnerable groups*". The European Commission, in its recommendation of October 2023, also points out:

> …addressing the root causes of energy poverty such as the low energy performance of homes and household appliances, high energy expenditure in proportion to household budgets and lower income levels (exacerbated by inflation). The recommendations are accompanied by a Staff Working Document which contains a more detailed analysis of the recommended measures. (EC, 2023b)

The Staff Working Document (EC, 2023c) provides detailed measures addressing the diagnosis, affordability, and underlying causes of energy poverty. We argue for the importance of translating these findings into regulatory measures at both EU and local levels.

Addressing energy poverty effectively requires implementing prescribed measures and allocating available funding where they are most needed. The broad strategy of general ERB programmes with high costs and lacking prioritisation criteria might not be the most effective approach. We urge establishing targeted measures alongside management mechanisms to optimise the allocation of limited funds. In Table 6.1, we contribute to this aim by proposing key indicators in the diagnostic process of VH required for effective ERB planning and implementation. Furthermore, the prevalence of energy poverty in social housing gives public authorities a key role in the optimisation of the ERB process.

Finally, we identify several problems concerning the accessibility and distribution of subsidies for ERB to the most vulnerable demographics. Firstly, there are important information deficits and management complexities. Secondly, there is an over-bureaucratisation of the

Table 6.1 Indicators of residential vulnerability to heat (VH)

Category	Heat vulnerability factor	Variables
Exposure	Ageing of buildings in relation to thermal insulation legislation applied	(**A**) % of dwellings built before 1950 (**B**) % of dwellings built between 1951 and 1980 (**C**) % of dwellings built between 1981 and 2007 (**D**) % of dwellings built after 2008
	Building quality	(**E**) Construction quality from 1 to 9 points
	Energy efficiency	(**F**) Energy demand (Kwh/m2 per year)
	Urban morphology	(**G**) Average height of buildings (m)
	Concentration/overcrowding	(**H**) % of dwellings with an area < 30m2 (**I**) Dwelling average surface (m2/dwelling) (**J**) Dwelling average surface per resident (m2/person)
Sensitivity	Loneliness/age	(**K**) % of population > 65 years living alone
	Gender/age	(**L**) % of women > 65 years
	Poverty	(**M**) Household average income (€/household per year) (**N**) Average income/person (€/person per year)
	Education	(**O**) % of population with higher education (**P**) % of population without studies
	Households	(**Q**) % of single-person households (**R**) % of single-parent households
	Foreign population	(**S**) % of non-EU foreign population

(continued)

Table 6.1 (continued)

Category	Heat vulnerability factor	Variables
Adaptation capacity	Potential of energy production	(**T**) Surface above-building per dwelling (m2/dwelling)
	Tenure status	(**U**) % of rented dwellings
	Water presence surface	(**V**) Area of swimming pools (m2/inhabitants)
	Urban greening	(**W**) Number of trees
		(**X**) Tree density in public spaces (trees/Ha.)
		(**Y**) Green space per capita (m2/person)
		(**Z**) Urban park surface area (m2/person)

Source Own elaboration based on literature review. See Appendix for further information

process exacerbated by cultural and digital gaps. Thirdly, there is limited public investment in integrated urban regeneration programmes, which require interdisciplinary teams and socio-educational initiatives. Moreover, the majority of funds are directed towards individual housing, as "the renovation of social and multi-apartment housing faces additional barriers due to the complex decision-making process" (EC, 2020, 22).

Heat Adaptation Actions: Temporal and Spatial Dimensions

Considering the above described problems, another challenge for heat adaptation policies is the prescription of temporal and spatial dimensions. First, a demarcation must be drawn between short- and medium-term measures. Short-term measures, urgent and swiftly implemented, demand proactive administrative actions. These actions include providing immediate, cost-effective solutions for vulnerable households and establishing networks of climatic shelters for extreme temperatures. Communication and awareness campaigns are also crucial. These actions could be promoted through networks of energy consultation points, which are known as "*energy coaching*" (Schneider et al., 2023), already operational in the Netherlands and the UK, and in some Spanish regions. Improving local and regional qualitative research on perception, awareness, and/or social alertness of heatwaves could also help an efficient enactment of short-term strategies. Medium-term measures include heat adaptation of public spaces and building retrofitting, largely underfunded at the moment. These measures require a pre-diagnostic phase to optimise resource allocation plus post-implementation auditing and evaluation.

Regarding the spatial dimension, defining suitable suburban scales for implementation requires the availability of reliable datasets for a proper diagnosis. The so-called "neighbourhood effect" (Aguado, 2021) is defined as the combination and accumulation of different factors that contribute to VH being aggravated by spatial processes of segregation and degradation. At the same time, the neighbourhood can also be perceived as an opportunity, since local identity and social relations create potential for transformation and resilience (Torres, 2021).

Macro-economic data overshadows escalating poverty rates in EU's vulnerable neighbourhoods. The EU's Gini coefficient stood at 30.1% in 2022, with the highest income inequality nearing 40%. In numerous EU countries, particularly in the Mediterranean and Eastern sub-regions, inequality is on the rise. Therefore, our recommendation is that ERB should be conceived as a mechanism for redistributing wealth and improving social cohesion. Special attention must be paid to socially vulnerable groups, including homeless and incarcerated individuals, and residents of informal settlements who have limited or no access to water and/or electricity supplies.

Cross-Cutting, Unified, and Comprehensive Databases for an Accurate VH Mapping

The paucity and inadequate quality of accessible data on the multiple factors of VH is a cross-cutting problem to the challenges described above. This deficiency stems from multiple reasons. First, census data is often discontinuous and of limited accuracy, particularly the data on the characteristics of housing construction. Second, there is a conspicuous absence of accurate data on energy consumption. Lastly, the reliability of heat-related mortality data is moderate and faces difficulties in cross-validation of official models. Addressing these problems requires a multifaceted approach by enhancing local catalogues of residential buildings, ensuring the availability of energy consumption at building level and the traceability of data on heat-related health problems.

Even with adequate data, the accurate mapping of VH requires a suitable operationalisation of the VH concept that, as suggested above, considers the intersection of biophysical and socio-economic factors in line with the framework of social determinants of health (Barton & Grant, 2006).

Multi-Criteria Assessment of Vulnerability to Heat

Table 6.1 describes an indicator set to assess the three dimensions of VH proposed by Wolf and McGregor (2013) and Domene et al. (2022): sensitivity, exposure, and adaptability. We propose this multi-criteria framework as a reference for prioritising ERB actions and funds. The choice of statistical indicators, based on a literature review, responds to a triple perspective: socio-economic, environmental, and related to the building characteristics. The integrative SSH-STEM approach within our team led to identifying and precisely adjusting key variables to each VH factor. In addition to these three dimensions, we suggest analysing the boundary conditions that intensify heat effects, such as the UHI phenomenon.

To test the proposed framework, we conducted a statistical multivariate analysis for the case study of Seville, taking the census section of the municipality as the minimum scale. We developed a matrix of 13,598 data, 26 variables for 523 census sections. The research followed an iterative process utilising Principal Component Analysis to reduce data dimensionality. This process influences the final selection of indicators crucial to the studied problem. The determinant of the Correlation Matrix, nearly zero, signifies appropriate data reduction, corroborated by Bartlett's test of sphericity results. The analysis suggests the possibility of creating synthetic indices by combining variables listed in Table 6.1. Depending on application sites, and considering the expressed variables, identifiable components of underlying issues can be obtained as guiding criteria for prioritisation of funds.

Our analysis reveals the following indicators as most prevalent in determining VH in Seville: income level; population over 65 (mostly elderly women living alone); educational level; ageing and quality of buildings; and urban greening. These results are coherent with similar multivariate analysis carried out in other Spanish cities such as Madrid or Barcelona (Domene et al., 2022; López-Bueno et al., 2020). Therefore, we suggest these indicators can be applied to cities with similar characteristics of residential and social VH situations. The present work contributes to identifying key VH factors and the specific variables of each. This methodology could also be used for developing technical databases to support preventive diagnostics in ERB programmes.

Conclusions and Recommendations

Identifying heatwave risks as a public health problem requires adopting a health-oriented approach to climate change policies, including ERB programmes. This means recognising heat as one of the EU's deadliest climate hazards and prioritising heat-related problems and measures, particularly in the Mediterranean region. Moreover, it involves viewing ERB programmes as adaptation strategies focused on the most vulnerable populations, within the framework of social determinants of health.

Given the high costs associated with ERB and the constraints of limited investments, accurate diagnostics are crucial to identifying priority actions and intervention areas. We propose a multi-criteria framework that can be used to prioritise ERB funds. Additionally, we identify a sub-set of indicators that best describe VH in Mediterranean cities: income level; population over 65 (mostly elderly women living alone); educational level; ageing and quality of buildings; and urban greening. To ensure reliable and rigorous results, it is vital to improve the quality and availability of databases concerning health, residential buildings, and energy consumption in dwellings.

Finally, the multidisciplinary dialogue at our expert seminar highlights the obstacles for vulnerable populations in accessing support for climate change adaptation due to information gaps and the complexities of managing overly bureaucratised processes. Addressing these barriers requires raising investment in ERB programmes with a comprehensive approach, interdisciplinary teams, and socio-educational initiatives such as networks of "energy coaching" points.

Altogether, the highlighted challenges and proposals in this chapter emphasise the need for interdisciplinary approaches both in research and implementation of climate change policies to ensure an efficient and just adaptation to increasing heatwave risks.

Appendix

Lara García, A., Cruz Mazo, E. C., Galán Marín, C., Núñez Rivera, C., & Rivera Gómez, C. (2024). *Vulnerability to heatwaves and public health: Identifying indicators for EU policies on energy renovation of residential buildings*. Zenodo. https://doi.org/10.5281/zenodo.11202289.

References

Aguado, I. (2021). El efecto barrio y su contestación en Bilbao. In O. Nello (Ed.). *Efecto barrio. Segregación residencial, desigualdad social y políticas urbanas en las grandes ciudades ibéricas* (pp. 360–397). Tirant lo Blanch.

Ballester, J., Quijal-Zamorano, M., Méndez Turrubiates, R. F., Pegenaute, F., Herrmann, F. R., Robine, J. M., Basagaña, X., Tonne, C., Antó, J. M., & Achebak, H. (2023). Heat-related mortality in Europe during the summer of 2022. *Nature Medicine, 29*, 1857–1866.

Barton, H., & Grant, M. A. (2006). Health map for the local human habitat. *Journal of the Royal Society for the Promotion of Health, 126*(6), 252–253.

Bouzarovski, S. (2014). Energy poverty in the European Union: Landscapes of vulnerability. *Wires Energy and Environment, 2014*(3), 276–289.

Croon, T. M., Hoekstra, J. S. C. M., Elsinga, M. G., Dalla Longa, F., & Mulder, P. (2023). Beyond headcount statistics: Exploring the utility of energy poverty gap indices in policy design. *Energy Policy, 177*(2023), 113579.

De la Osa, J. (2016). *Cambio climático y salud. Actuando frente al cambio climático para mejorar la salud de las personas y del planeta*. Observatorio de Salud y Medio Ambiente. DKV Seguros. ECODES.

Domene, E., Lacort, E., & Saavedra, B. (Coord.). (2022). *El calor en un futuro: Índice de vulnerabilidad al cambio climático (IVAC)*. Institut Metròpoli.

European Commission. (2020). *A renovation wave for Europe—Greening our buildings, creating jobs, improving lives*. Publications Office of the European Union.

European Commission. (2023a). *EU-level technical guidance on adapting buildings to climate change*. Publications Office of the European Union.

European Commission. (2023b). *Energy, climate change, environment. Commission publishes recommendations to tackle energy poverty across the EU*. Publications Office of the European Union.

European Commission. (2023c). *Commission staff working document EU. Guidance on energy poverty accompanying the document commission recommendation on energy poverty*.

European Commission. (2024, April). *Renovation wave. Aiming to improve energy efficiency, boost the economy and deliver better living-standards for*

Europeans. European Commission. Directorate-General for Energy. https://energy.ec.europa.eu/topics/energy-efficiency/energy-efficient-buildings/renovation-wave_en#main-focus-of-the-renovation-wave

Eurostat. (2024, February). *Population structure and ageing*. https://ec.europa.eu/eurostat/statistics-explained/index.php?title=Population_structure_and_ageing

Hernández, A., Rodríguez, I., & Córdoba, R. (Coord.). (2018). *Barrios vulnerables de las grandes ciudades españolas. 1991/2001/2011*. Instituto Juan Herrera.

IPCC. (2022). *Climate change 2022: Impacts, adaptation and vulnerability*. Sixth Assessment Report of the IPCC. Cambridge University Press.

López-Bueno, J. A., Díaz, J., Sánchez-Guevara, C., Sánchez-Martínez, G., Franco, M., Gullón, P., Núñez, M., Valero, I., & Linares, C. (2020). The impact of heat waves on daily mortality in districts in Madrid: The effect of sociodemographic factors. *Environmental Research, 190*, 109993.

Schneider, P. T., van de Rijt, A., Boele, C., & Buskens, V. (2023). Are visits of Dutch energy coach volunteers associated with a reduction in gas and electricity consumption? *Energy Efficiency, 16*, 42. https://doi.org/10.1007/s12053-023-10116-6

Torres, F. J. (2021). Polígono Sur en Sevilla. Historia de una marginación urbana y social. *Scripta Nova, 25*(2).

Wolf, T., & McGregor, G. (2013). The development of a heatwave vulnerability index for London, United Kingdom. *Weather and Climate Extremes, 1*, 59–68.

World Health Organisation. (2023). *WHO review of health in Nationally Determined Contributions and long-term strategies: Health at the heart of the Paris Agreement*. World Health Organization.

Open Access This chapter is licensed under the terms of the Creative Commons Attribution 4.0 International License (http://creativecommons.org/licenses/by/4.0/), which permits use, sharing, adaptation, distribution and reproduction in any medium or format, as long as you give appropriate credit to the original author(s) and the source, provide a link to the Creative Commons license and indicate if changes were made.

The images or other third party material in this chapter are included in the chapter's Creative Commons license, unless indicated otherwise in a credit line to the material. If material is not included in the chapter's Creative Commons license and your intended use is not permitted by statutory regulation or exceeds the permitted use, you will need to obtain permission directly from the copyright holder.

CHAPTER 7

Reforming Carbon Accounting Mechanisms Around Justice-Based Principles to Promote Societal Sustainability

Camilla Seeland, Piers Reilly, Ilaria Perissi, Diego Andreucci, Roger Samsó, and Jordi Solé

Policy Highlights To achieve the recommendation stated in the title, we propose the following:

- Societal benefit must replace the current economic-based rationale as the grounds for justifying proactive, targeted, and equitable carbon-reducing interventions in industry sectors.
- The policy priorities of the EU Commission and EU Parliamentary Committees must transition away from a focus on GDP expansion, and move towards the promotion of societal sustainability within the EU.

C. Seeland (✉) · P. Reilly · I. Perissi
Global Sustainability Institute and Faculty of Business and Law, Anglia Ruskin University, Cambridge, UK

D. Andreucci
Department of Geography, Faculty of Geography and History, University of Barcelona, Barcelona, Spain

© The Author(s) 2024
E. Galende Sánchez et al. (eds.), *Strengthening European Climate Policy*, https://doi.org/10.1007/978-3-031-72055-0_7

- To design policies driven by societal sustainability principles, the opportunity to develop a strong evidence base must be available.
- The carbon budget should be equitably divided among industry sectors based on outputs from the developed index.
- The inclusion of mixed data, qualitative and quantitative, should be considered across all policy areas to provide richer evidence bases and ensure robust policymaking.

Keywords Societal benefit · EU Emissions Trading System · Carbon credits · Fit for 55 · Carbon use efficiency

Introduction

The biggest problem facing the EU due to climate change is instability (EEA, 2024). The existing carbon credits system, the EU Emissions Trading System (EU ETS), while successful, has two fundamental flaws: it is not comprehensive enough (European Commission, 2024), and it is not equitable (Ellerman & Joskow, 2008). To maintain pace with competing global players, the EU must address the issue of climate change due to its status as a threat multiplier, which exacerbates current and future vulnerabilities and creates instability (Goodman & Baudu, 2023). All resources are limited, time, money, commitment, etc., therefore interventions must be carefully selected, with strong evidence bases to support those decisions.

The EU's Fit for 55 package was designed with the aim of reducing Greenhouse gas (GHG) emissions by at least 55% by 2030 (European Council, 2024). The process has guiding proposals that aim to ensure that: (1) the transition is socially fair and just, (2) the competitiveness of the EU is maintained and strengthened, and (3) the EU should be

---

R. Samsó · J. Solé
Centre for Ecological Research and Forestry Applications (CREAF), Cerdanyola del Vallès, Spain

Departament de Dinàmica de la Terra i de l'Oceà, Facultat de Ciències de la Terra, Universitat de Barcelona, Barcelona, Spain

a global leader in the fight against climate change (European Council, 2024). The Emissions Trading System (EU ETS) was introduced to allow cost-effective net-zero reductions (IPCC AR6, 2022), to take place at a balanced and manageable pace, with free allocation to key industries (European Commission, 2024). This was the world's first, and as of writing, remains the world's largest carbon market.

Under COP26 (UNFCCC, 2024, Glasgow), the foundation for an international carbon trading market was laid, allowing countries to exchange carbon credits issued by the United Nations to fulfil GHG reduction commitments as per the Paris Climate Agreement. These credits are aimed at providing a way to offset GHG emissions via projects and initiatives such as deforestation prevention, tree planting, and soil management. Voluntary markets predominantly support carbon offsetting and decarbonisation programmes of traditionally heavy-polluting industries like the fossil fuel, aviation, and technology sectors striving for net-zero emissions.

However, recent investigations into carbon offsetting protection programmes raise concerns with the global carbon market (West et al., 2023). This includes major problems with transparency in carbon accounting, leading to uncertainty of the effectiveness of these programmes (Perissi & Jones, 2022). A global carbon market is an unrealistic mechanism due to the required scale and lack of control opportunities. Regional markets are where success can be obtained, however, revisions are needed to capture their full potential, particularly the EU ETS.

As of November 2023, the EU Council and Parliament have agreed to revise the Directive on Industrial Emissions (IED) and the Industrial Emissions Portal (IEP) (European Council, 2023), which aims to "better protect human health and the environment by reducing emissions from certain industrial installations" including livestock, mining, energy, and aviation by enacting a polluter pays principle through improved emissions reporting (European Council, 2023). This is a necessary and admirable step which equitably balances the damage heavy polluters cause, but it does not go far enough. Additionally, the IED and IEP frameworks are based on Environmental Performance Limit Values (EPLVs), which are only binding for energy resources. Therefore, EPLVs are incomplete because they fail to consider requirements for life and fundamental pillars of society.

The greatest return on investment can be obtained via proactive targeted interventions, developed by redefining a successful transition to a low-carbon future as one based on strong societal sustainability, which includes "both positive and negative impacts of systems, processes, organisations, and activities on people and social life (i.e., health, social equity, human rights, labour rights, practices and decent working conditions, social responsibility, justice, wellbeing, etc.)" (Balaman, 2018, p. 86). This requires understanding that the low-carbon transition needs to go beyond energy considerations and instead requires a more comprehensive understanding of industry sectors outputs and how they impact society.

The key factor in successfully reducing a region's overall emissions is to account for all industries, not just the heaviest polluters based on scope 1 emissions, which are the direct emissions produced, owned, and controlled by a company; but also scope 2 and 3 emissions, the indirect emissions that come from purchased energy and the wider supply chain. These must be factored into carbon accounting, with a shift away from using economic metrics as primary tools for benchmarking. Free market principles, unfortunately, under the current system, do not act fast or thoroughly enough to achieve the emissions reductions required.

To overcome these challenges, it is crucial to look beyond climate change as a matter of economics and natural science: success requires moving towards comprehensive policy creation. A foundational aspect of this is to enhance collaboration between STEM and SSH research to create continuous multidisciplinary collaboration. Current policymaking models primarily focus on physical and technical datasets, failing to account for the importance of societal factors. Thus, strengthening the synergy between STEM and SSH disciplines is essential for driving comprehensive and effective policy design, which ensures that both technical and societal aspects are considered in decision-making processes.

Our team, two SSH researchers and three STEM researchers set the basis for developing a way to jointly assess physical and societal impacts of industry sectors. The goal of blending these datasets was to allow policymakers to identify sectors and target interventions to support transition pathways. To achieve this, we assessed the available physical and social datasets relevant to the EU and have developed an index that can be used across the EU. The index has been developed via participatory methods and is rooted in quantitative–qualitative methods. This index provides policy makers with defensible actions that are replicable and scalable, demonstrating the viability of proposed policy strategies. Our innovative approach prioritises carbon use efficiency, going beyond a simple reduction in emissions and moving towards maximising overall positive societal outcomes.

This research revealed that existing datasets relating to social factors are not sufficient to create strong, evidence-based policies for carbon emission reductions. Our attempted construction of a societal-benefit index revealed that it is currently not possible to create a comprehensive index due to the few existing social datasets available (2 usable social datasets) compared to the physical (98 physical datasets). The EU would greatly benefit from the gathering of additional social datasets to understand the links between environmental impact, societal benefit, and carbon emission reductions. Although the physical datasets are useful, they lack depth, which can be mitigated through the inclusion of social data.

Our methodology was composed of two parts. Part one was concerned with the creation of an index that portrays trade-offs between physical impacts and social impacts of economic sectors within the EU. To populate the index, we used the EXIOBASE3 database (Stadler et al., 2021), which offers a chronological sequence of Environmentally Extended Multi-Regional Input–Output (EE-MRIO) tables spanning from 1995 to 2022. EE-MRIO tables were used to integrate input–output analysis with environmental data to assess the environmental impacts of economic activities across regions and sectors. This data allowed us to measure environmental impacts and societal benefit added by different economic sectors within the EU and to create a ranking system (see Appendix).

Part two explored potential future datasets by engaging academics, businesses, and community interest groups, in a workshop developed via participatory methods. The workshop gathered consensus on which social needs were crucial and why, with the outcomes revealing a ranking of desirable social needs that we used to develop our societal-benefit framework. This exposed detrimental gaps in EU-level data gathering, which while exemplary for economic and natural science-based criteria, wholly lacks any formal attempt to gather and centralise social data, reducing the effectiveness, or utterly preventing, evidence-based policymaking (see Appendix).

The Evidence Base

Our research shows that in-depth knowledge and understanding of industry-based societal benefit impacts is a requirement for designing robust policy. Existing datasets do not contain sufficient levels of information and so evidence bases for Just, future-proof policies cannot be developed. Our exploration in designing this societal-benefit index initially found just 28 potential social datasets that could be used. After further evaluation, only 2 of the 28 potentials were suitable. This is in comparison to the 98 physical data sets, all of which were usable. This was due to the differences in how the data sets were presented, preventing direct comparison.

This means policy makers cannot be provided with evidence of industries' social outputs on the same level as their physical outputs, leaving a gap in understanding of how interventions in industries should be carried out, and the impacts those interventions could have.

Table 7.1 shows workshop participants stated needs from most important (1) to less important (4) based on two different scenarios. "Present day" asked them about their current needs based on the contemporary environmental situation. In terms of current needs, we made no distinction between what participants would classify as personal and societal needs, for our purposes those two categories are a needless distinction. The "SSP3-7.0 Regional Rivalry" scenario asked participants to list their needs based on an environment where "emissions and temperatures rise steadily and CO_2 emissions roughly double from current levels by 2100. Countries become more competitive with one another, shifting toward national security, and ensuring their own food supplies. By the end of the century, average temperatures have risen by 3.6 °C" (IPCC AR6, 2022).

Our workshop helped us identify desired social datasets. Our workshop participants stated that the social datasets should be expanded to include education (i.e., education at all levels and ages), health (i.e., physical and mental health and well-being), social equality (i.e., provisioning for vulnerable, balance of income and living cost, inclusive welfare system), human thriving (i.e., employment job quality, unemployment, non-employment) and ecosystem thriving (i.e., nature-centric designations) datasets. When workshop participants were asked to list their everyday "needs", food, shelter, and energy were the main priorities, followed by health, community, knowledge, and a purpose in life. When the same question was asked under the IPCC SSP3-7.0 regional rivalry

Table 7.1 Workshop participants' collective decisions and suggested social datasets

	Present day	*SSP3-7.0: regional rivalry*
1	Food, shelter, energy, security	Food, shelter, water, energy, security, land
2	Health, well-being, community (*sense of belonging, family, and inclusion*)	**Community** (*loved ones*), **inclusivity** (*Local community*), **rules** (*order*)
3	Knowledge, purpose, education	**Health, well-being**
4	Leisure, travel, creativity, nature, Wi-fi	**Skills, awareness/connectedness**

Suggested social datasets: Education, health, inequality, social mobility, well-being, ecosystem health, quality of life, human thriving

scenario, priorities were similar, however, security and land became the top priorities. Due to the ranking of their needs collectively, debate coalesced around the individual vs society without intervention from the facilitator or convenor of the workshop. The list of needs stated by our participants informed our proposed social datasets. Furthermore, to better understand societal benefit, policymakers need to engage more interactively with society; workshop participants requested both traditional and dynamic methods of engagement and information sharing, providing examples of short documentaries, animated cartoons, and infographics, alongside more traditional executive summaries.

Conclusion and Recommendations

Our research led us to one overall policy recommendation, allowing new benchmarking and longer-term goal setting to take place. This will provide greater clarity, transparency, and actionability for all sectors of society. This overall recommendation is accompanied by a four-step plan which can be acted upon by different EU-level bodies creating a manageable route map to success: **Societal benefit must replace the current economic and monetary-based rationale as the grounds for justifying proactive, targeted, and carbon-reducing interventions in industry sectors.**

Step 1: The policy priorities of the EU Commission and EU Parliamentary Committees must transition away from a focus on GDP expansion, and move towards the promotion of societal sustainability within the EU.

The dependence on primarily financial market-based mechanisms for climate mitigation does not guarantee, in fact it jeopardises, the achievement of objectives outlined in the Fit for 55 package. When assessing the commitment of industry sectors to climate action, it is imperative to transcend financial capacity for mitigation and consider social output of industries through a societal-benefit lens. This approach entails the integration of social dimensions into the low-carbon transition, recognising that while there may be an initial drawback, namely a mild deceleration of decarbonisation progress, it opens avenues for expediting rapid transition in the near future.

External, international projects should continue to be explored. However, due to the lack of direct control the EU and member states can exert upon carbon markets and mechanisms beyond the EU border, efforts should be focused domestically. This includes processes that intersect with the EU market, such as further development of the carbon border and new carbon import standards. Thus, policymakers are urged to consider these intricacies as they formulate strategies and policies for effective, sustainable, and future focused climate mitigation efforts.

Step 2: To design policies driven by societal sustainability principles, the opportunity to develop a strong evidence base must be available.

Data is crucial for evidence-based policymaking. Our research is an example of coordinating SSH-based research with STEM-based research to design comprehensive policies for a sustainable future. To do this, the data collection of qualitative metrics needs to be organised, mirroring the existing quantitative datasets as far as possible. This data coordination would not require a heavy investment since much of this data is already being gathered by member states, it merely needs to be collated. We call for a unified policy design approach, created at the EU level, to ensure that sector-by-sector analysis can occur to benefit the whole of society.

Step 3: The carbon budget should be equitably divided among industry sectors based on outputs of the developed index.

Under the current EU ETS, carbon credits can be sold, bought, and traded. This has resulted in carbon credits being treated as a commodity and source of economic revenue. This neither motivates nor assists industries with their transition to lower carbon practices and risks increasing the potential for greenwashing and double-counting emission reductions (López-Vallejo, 2021). For the EU to effectively and accurately reduce overall emissions, the trading of carbon credits needs to be discontinued. The allocation of carbon allowances and credits needs to be centralised within the ETS and the EU Commission, whose role it will be to divide the annually decreasing carbon allowances and credits based on an industry's overall societal benefit.

We advocate that those sectors with a high societal benefit, e.g., health industry, agriculture, water treatment, etc., should continue to be protected via free carbon allowances and supported in their decarbonisation journey. Sectors that provide no or low societal benefit should be restricted from free credit allocation and obligated to transition or divest. Industry sectors are working hard on reducing their Scope 1 emissions, however achieving net-zero requires tackling Scope 2 and 3 emissions, which are vital to understanding gross supply chain emissions (National Grid Group, 2023).

We propose an expanded EU-wide carbon credits system, allocated by the ETS and EU Commission, to support industries providing a high societal benefit. We anticipate this will (1) speed up the transition as the private sector is encouraged to transition to net-zero, (2) prioritise and support industry sectors critical for societal well-being, (3) provide citizens, businesses, and member states with certainty to help their long-term decision-making, and (4) increase stability within the EU economically, socially, and environmentally.

Step 4: The inclusion of mixed data, qualitative and quantitative, should be considered across all policy areas to provide richer evidence bases and ensure robust policymaking.

During our research, while actively engaging and communicating to bridge SSH and STEM disciplines, we discovered the added richness, depth, and understanding that came from collaboration and interdisciplinary research. We strongly encourage further explorations of mixed teams such as ours to create shared ground between the natural and social sciences. This should be replicated across other policy areas which we believe is the key to managing current and future EU-level problems.

Overall, we suggest that while the EU ETS is an excellent step in the right direction, it does not currently go far enough. By basing the carbon credit mechanisms on a purely economic foundation, opportunities for greater action have been missed. This can be rectified by replacing the current economic-based rationale with societal-benefit considerations, allowing policy makers to develop proactive, targeted, and equitable carbon-reducing interventions. We hope that by adopting our policy recommendations, rapid success can be achieved in designing these interventions. Furthermore, we hope that this helps create a more comprehensive decarbonisation plan that demonstrates a more prosperous, competitive, and sustainable future can be achieved.

Acknowledgements This chapter is the culmination of the work of many more people than the named authors. We would like to thank the SSH Centre Team, in particular Dr Ami Crowther, Professor Chris Foulds, and Professor Rosie Robison, with further thanks to the Horizon Europe Research and Innovation Programme, and UK Research and Innovation. We would like to thank our editor Sara Heidenreich and our reviewers for their support and valuable input. We would also like to thank Enric Alcover Comas for their hard work and problem-solving abilities. We would like to thank Anglia Ruskin University, with special thanks to the Global Sustainability Institute and the Faculty of Business and Law for their support, as well as the University of Barcelona, the Marie Curie Fellowship for enabling us to conduct this interdisciplinary research. Finally, a massive thank you to our workshop participants, without whom this research would not have been possible.

Appendix

Seeland, C., Reilly, P., Perissi, I., Andreucci, D., Samsó, R., & Solé, J. (2024). *Reforming carbon accounting mechanisms around justice-based principles to promote societal sustainability*. Zenodo. https://doi.org/10.5281/zenodo.11384376.

REFERENCES

Balaman, S. Y. (2018, October 2). *Decision-making for biomass-based production chains.* Academic Press (Accessed 01 May 2024, 11:00).

Ellerman, A. D., & Joskow, P. L. (2008). *The European Union's emissions trading system in perspective.* Pew Center on Global Climate Change. https://economics.mit.edu/sites/default/files/2022-09/EU%20Emissions%20Trading%20System%20in%20Perspective.pdf (Accessed 01 May 2024, 11:00).

European Commission. (2024). *Ensuring the integrity of the European carbon market.* Climate.ec.europa.eu. https://climate.ec.europa.eu/eu-action/eu-emissions-trading-system-eu-ets/ensuring-integrity-european-carbon-market_en (Accessed 01 May 2024, 11:00).

European Council. (2023). *Industrial emissions: Council and Parliament agree on new rules to reduce harmful emissions from industry and improve public access to information.* European Council. https://www.consilium.europa.eu/en/press/press-releases/2023/11/29/industrial-emissions-council-and-parliament-agree-on-new-rules-to-reduce-harmful-emissions-from-industry-and-improve-public-access-to-information/ (Accessed 01 May 2024, 11:00).

European Council. (2024). *Fit for 55—The EU's plan for a green transition—Consilium.* European Council. https://www.consilium.europa.eu/en/policies/green-deal/fit-for-55-the-eu-plan-for-a-green-transition/ (Accessed 01 May 2024, 11:00).

European Environment Agency. (2024). *Climate change impacts, risks and adaptation.* www.eea.europa.eu. https://www.eea.europa.eu/en/topics/in-depth/climate-change-impacts-risks-and-adaptation (Accessed 01 May 2024, 11:00).

Goodman, S., & Baudu, P. (2023). *Climate change as a 'threat multiplier': History, uses and future of the concept.* Centre for Climate and Security Briefer, 38. https://climateandsecurity.org/2023/01/briefer-climate-change-as-a-threat-multiplier-history-uses-and-future-of-the-concept/ (Accessed 01 May 2024, 11:00).

IPCC. (2022). *Sixth Assessment Report, AR6—IPCC.* IPCC. https://www.ipcc.ch/assessment-report/ar6/ (Accessed 01 May 2024, 11:00).

López-Vallejo, M. (2021). Non-additionality, overestimation of supply, and double counting in offset programs: Insight for the Mexican carbon market. In *Towards an emissions trading system in Mexico: Rationale, design and connections with the global climate agenda.* Springer Climate (pp. 191–221). Springer.

National Grid Group. (2023). *What are scope 1, 2 and 3 carbon emissions?* https://www.nationalgrid.com/stories/energy-explained/what-are-scope-1-2-3-carbon-emissions#:~:text=As%20the%20Greenhouse%20Gas%20Protocol,on%20the%20greatest%20reduction%20opportunities%E2%80%9D (Accessed 01 May 2024, 11:00).

Perissi, I., & Jones, A. (2022). Investigating European Union decarbonization strategies: Evaluating the pathway to carbon neutrality by 2050. *Sustainability, 14*, 4728.

Stadler, K., Wood, R., Bulavskaya, T., Södersten, C.-J., Simas, M., Schmidt, S., Usubiaga, A., Acosta-Fernández, J., Kuenen, J., Bruckner, M., Giljum, S., Lutter, S., Merciai, S., Schmidt, J. H., Theurl, M. C., Plutzar, C., Kastner, T., Eisenmenger, N., Erb, K.-H., et al. (2021). *EXIOBASE 3 (3.8.2)*. Zenodo. https://doi.org/10.5281/zenodo.5589597

UNFCCC. (2024). *Outcomes of the Glasgow Climate Change Conference.* https://unfccc.int/process-and-meetings/conferences/glasgow-climate-change-conference-october-november-2021/outcomes-of-the-glasgow-climate-change-conference (Accessed 01 May 2024, 11:00).

West, T. A. P., Wunder, S., Sills, E. O., Börner, J., Rifai, S. W., Neidermeier, A. N., Frey, G. P., & Kontoleon, A. (2023). Action needed to make carbon offsets from forest conservation work for climate change mitigation. *Science, 381*(6660), 873–877.

Open Access This chapter is licensed under the terms of the Creative Commons Attribution 4.0 International License (http://creativecommons.org/licenses/by/4.0/), which permits use, sharing, adaptation, distribution and reproduction in any medium or format, as long as you give appropriate credit to the original author(s) and the source, provide a link to the Creative Commons license and indicate if changes were made.

The images or other third party material in this chapter are included in the chapter's Creative Commons license, unless indicated otherwise in a credit line to the material. If material is not included in the chapter's Creative Commons license and your intended use is not permitted by statutory regulation or exceeds the permitted use, you will need to obtain permission directly from the copyright holder.

CHAPTER 8

Leave No One Behind: Engaging Communities in the Just Transition Process Towards Climate Neutrality

Ricardo García-Mira, Nachatter Singh Garha, Serafeim Michas, Franziska Mey, Samyajit Basu, and Diana Süsser

Policy Highlights To achieve the recommendation stated in the title, we propose the following:

- Equip key actors with the transformative capacities to support the development and implementation of regional visions, plans, and narratives.

R. García-Mira (✉) · N. S. Garha
University of A Coruna, A Coruña, Spain
e-mail: ricardo.garcia.mira@udc.es

N. S. Garha
e-mail: nachatter.singh@udc.es

S. Michas
University of Piraeus, Pireas, Greece
e-mail: michas@unipi.gr

© The Author(s) 2024
E. Galende Sánchez et al. (eds.), *Strengthening European Climate Policy*, https://doi.org/10.1007/978-3-031-72055-0_8

- Develop structures and participatory mechanisms that encourage a wide social dialogue and citizen involvement in just transition projects.
- Use insights from science and practice from participatory processes, as well as methods and tools developed in different contexts, to enable just transitions.
- Involve local stakeholders in the intersectional analysis of compensatory measures for holistically mitigating negative impacts of policies or interventions.
- Make use of the full spectrum of SSH and STEM tools to support local transition processes.

Keywords Just transition · Community engagement · Stakeholder participation · Just transition plans · Socio-economic impact

Introduction

Phasing out fossil fuels in a just manner is essential for achieving EU climate objectives. Currently, the closing down of fossil fuel industries and the transition towards a renewable energy system in Europe and beyond are taking place unevenly, with very different decarbonisation strategies. This is partly caused by regions having different levels of dependence on fossil fuels, transformative capacities, and the availability of

---

F. Mey
Research Institute for Sustainability, Helmholtz Centre Potsdam, Potsdam, Germany
e-mail: franziska.mey@rifs-potsdam.de

S. Basu
Mobilise Mobility and Logistics Research Group, House of Sustainable Transitions, Vrije Universiteit Brussel, Ixelles, Belgium
e-mail: samyajit.basu@vub.be

D. Süsser
Institute for European Energy and Climate Policy, Amsterdam, The Netherlands
e-mail: diana@ieecp.org

human and natural resources. As a result, the way the energy transition locally unfolds has a diversified agenda, which means there is no silver bullet for all coal- and carbon-intensive regions (CCIRs). The term "just transition" is high on the political agenda (Lee & Baumgartner, 2022), and an increasing number of countries have developed transition plans. However, the policy interpretation of a just transition varies from country to country (Hermwille et al., 2023). The debate is generally moving towards a more holistic understanding of the transition to a more sustainable energy system. This involves taking into account the environmental, social, demographic, and economic impacts of the transition on all members of society (Abram et al., 2022; Wang & Lo, 2021).

The European Green Deal aims to leave no one behind in the transition, although implementing this goal in practice is not exempt from challenges. In this sense, the Just Transition Mechanism is a crucial mechanism that promotes social and economic justice and enables structural change in the regions most affected. It consists of three pillars: (1) the Just Transition Fund; (2) a dedicated scheme under the InvestEU package, and (3) a new public sector loan facility. To access the different funds, eligible regions across Europe must submit their Territorial Just Transition Plans (TJTP) to the European Commission (European Commission, n.d.). However, despite these efforts, some authors have warned that the European Green Deal policies fail to address the social dimension and thus might exacerbate social inequalities (Akgüç et al., 2022). When it comes to implementation, meaningful engagement will again be critical for giving citizens and communities ownership of the process. In this book chapter, we highlight common overarching challenges that need to be addressed to achieve just transitions: the persistence of top-down approaches from the EU or national level in formulating just transition plans, the inadequate representation of affected communities and stakeholders in developing strategies and solutions for their own regions, and the lack of exploratory assessment of solutions considering diverse contexts and place-based challenges. Finally, we draw recommendations for meaningful stakeholder and community engagement in transition processes to address these challenges.

This chapter draws on SSH and energy modelling applied in four European projects: TIPPING+, JUSTEM, ENTRANCES, and TANDEM. Researchers from different fields, such as psychology, sociology, economics, geography, political science, energy research, and demography, participated in two internal workshops to identify overarching challenges, approaches, and solutions to enable place-based and citizen-centred transitions. The results from these projects show that science can provide important insights and methods and tools, including scenarios, for policymaking to enable just transitions.

COMMUNITY ENGAGEMENT FOR A JUST TRANSITION

Scientific evidence underscores the pressing need for achieving climate neutrality by 2050 while also emphasising the necessity for societies to benefit from the transition process, thereby requiring active engagement from affected communities (Devine-Wright, 2007; Schot et al., 2016; Sovacool et al., 2022). However, a better understanding of different social groups' barriers and the impact of engagement strategies on them is essential (Hanke et al., 2021).

Stakeholder Engagement Challenges and Opportunities

This section provides insights into the opportunities and challenges of community engagement from four European projects. Evidence from the JUSTEM project indicates large differences in how citizens and stakeholders have been engaged in preparing TJTPs in coal regions (Koasidis et al., 2023). For example, in Poland and Spain, stakeholder engagement followed a more structured approach to ensure the involvement of a large number of stakeholders. Romania has employed multi-level stakeholder participation and set up a stakeholder registry with around 290 identified organisations and/or persons, which helped to involve local stakeholders. In Croatia and Bulgaria, regional authorities had a minimal role in preparing the TJTPs. In Greece and Bulgaria, whether comments from public consultations were considered is not evident. Overall, citizens have been insufficiently involved in drafting transition plans in all studied regions. Despite the very active discussions and the great need to be listened to expressed by participants, experiences from the JUSTEM project also revealed challenges in gaining the interest of citizens in participating in workshops (KAPE et al., 2023). In any case, there has been a lack of transparency on how stakeholder views and contributions have been incorporated into the plans (Koasidis et al., 2023). This fact highlights the need for new ways of meaningful and continuous collaboration between national and regional authorities that truly value and consider local knowledge, concerns, and aspirations.

The TANDEM project identified challenges regarding the involvement of vulnerable groups in the transition initiatives in CCIRs. Based on an analysis of 27 initiatives, the project showed that one-third of these initiatives did not identify or involve vulnerable groups associated with the specific actions. Only 13 out of 27 initiatives promote the participation of vulnerable groups during some parts of the implementation phase. This resulted in undesirable effects, such as the deepening of inequalities or the inaccessibility of compensatory measures to specific groups (e.g., tenants, low-income groups), and direct negative effects (e.g., an increase in energy prices following renovations or the installation of renewable energy sources).

In the ENTRANCES project, the approach to citizen involvement was thoroughly refined, and various methods, such as surveys, focus groups, stakeholder interviews, and co-creation meetings, drawing on SSH and STEM knowledge, were tested and implemented, demonstrating their effectiveness. To this purpose, a comprehensive framework of knowledge co-creation for the CCIRs in transition was developed. It specifically considered the views, opinions, representations, and tacit knowledge of a variety of stakeholders at local, regional, European, and international levels who were involved in the research process and in the production of the project's recommendations. This co-creation process consisted of a fact-based dialogue involving representatives of the Quintuple Helix, i.e., researchers, policymakers, businesses and industry, citizens, and defenders of the natural environment (Carayannis & Campbell, 2010). The dialogue was supported by the different results of the project: regional case studies, a taxonomy of challenges and coping strategies, socio-ecological and technical scenarios, and socio-economic simulations. Three subsequent meetings were conducted. The objective of these meetings was to gather insights on the strategies and approaches adopted or planned for regional development in each region, as well as on the main barriers and facilitators from the perspectives of the different stakeholders. In addition to the co-creation meetings, thirteen focus groups (one in each CCIR) were conducted to collect qualitative data and co-create knowledge on the socio-cultural component of the multidimensional analytical framework used to study the social impacts of decarbonisation and energy transition processes in the CCIRs. The semi-structured interviews with local stakeholders, considered key informants of the research, were conducted to collect primary data as well as the viewpoints of the different stakeholders on the socio-ecological and socio-technical (SETS) aspects of the energy transition in the different CCIRs (Garha & Garcia Mira, 2023). These meetings helped researchers assess the transformative capacities (Strasser et al., 2019) of the CCIRs in transition, which refer to the ability of these areas to adapt, innovate, and transition towards more sustainable and less carbon-intensive economic activities, and contributed to the formulation of policy recommendations.

In TIPPING+, the researchers also applied various methods, such as desk research, stakeholder interviews, workshops, and energy modelling, to explore just transition pathways. By iteratively combining these methods, stakeholder engagement was facilitated and led to the identification of positive transition narratives in selected communities, which deviate from not-so-just marginal transition pathways. One of the case studies, the City of Megalopolis in Greece, highlighted how the integration of energy modelling in stakeholder engagement strategies could support the creation of transition pathways with direct benefits for the citizens. Megalopolis is a city whose workforce has largely been employed in lignite mining and coal-based power generation since the 1970s. Following national mandates, lignite power plants in the city will be shut down, leading to the loss of jobs and the loss of the lignite-fuelled district heating network, which supplied cheap space heating for the households of Megalopolis. To counter the loss of district heating, a natural gas distribution network is being developed, and all households in Megalopolis are being supplied with a new natural gas boiler free of charge. The local community embraced this decision, as the cost of purchasing a new heating boiler was avoided, and natural gas was still cheap at the time of the decision. To initiate discussions on the topic, a first "get-to-know" stakeholder workshop was held, where participants shared their perspectives. The discussions led to the elaboration of alternative heating transition scenarios for the residential sector of Megalopolis, where instead of using natural gas as an intermediate fuel, heat pumps are deployed from the beginning. To quantify the effect of such scenarios, the DREEM model (Stavrakas & Flamos, 2020) was used to compare the energy consumption, environmental footprint, and potential extra charge on households for residential heating. The results of the DREEM model were presented during a field visit to Megalopolis, during which the findings were discussed with the city mayor, the local energy-producing company, academic representatives, and a non-profit organisation. The discussion led to the idea of transforming Megalopolis into a green city by prioritising renewable energy from the 550 MW of planned photovoltaics in the region for direct consumption in the city with the aid of batteries, maximising the synergies with heat pumps. To quantify such a scenario, the STREEM model (Michas & Flamos, 2023) was used to calculate the required renewable energy production to cover 90% of the city's demand with green energy, the accompanying storage capacity to enable this, and the cost of renewable energy supply, comparing it to the cost of purchasing electricity from the grid. The analysis showed that covering 90% of the city's electricity demand with direct solar energy and using heat pumps instead of natural gas for heating could save households up to 1700 euros per year compared to the current transition scenario. The relevant results were presented at the second and final stakeholder workshop, where participants were able to understand the trade-offs and long-term benefits of accelerated green transition pathways compared to using fossil fuels as intermediate solutions. The entire process highlighted that using energy modelling and communicating the results to stakeholders in an easily digestible manner can enhance the common understanding among various stakeholder groups towards long-term sustainable transition pathways.

Cross-Project Learnings on the Added Value of Community Engagement

Building upon the insights from the preceding section, we identified challenges and opportunities for just transition implementation processes. The evidence from the different EU projects suggests persistent issues with local stakeholder and community engagement in the systematic transition of CCIRs, specifically in developing future strategies and plans such as the TJTP. Our examples emphasise that this can lead to undesired outcomes when, for example, policy measures do not align with local needs and inequalities deepen. Consequently, a lack of local buy-in may fuel resistance and social tension. This has been particularly evident in the JUSTEM and TANDEM projects, emphasising the need to involve local and regional stakeholders and citizens across the entire policy process, from policy development through implementation. To achieve this goal, we find that stakeholders and citizens must be equipped with the necessary knowledge, skills, and funding to fulfil their roles and strengthen their transformative capacities to support and own the transition process in their regions. Multiplicity and interdependence of vulnerability factors may result in unintentional overlooking of social groups or individuals in policy decisions (e.g., negative effects of the closure of a thermal power station on women from lower socio-economic groups involved in maintenance or cleaning jobs or in a staff canteen). Therefore, the projects highlighted the need to involve local stakeholders and community members in the intersectional analysis of compensatory measures for holistically mitigating the negative impacts of the policy intervention.

Projects like ENTRANCES and TIPPING+ have set out to test and apply methods for stakeholder and citizen participation in addressing these issues in different engagement formats. Learnings from these projects confirmed the added value and importance of local voices for, on the one hand, a better understanding of the transition process in CCIRs and, on the other hand, showcasing how to empower community members to express their needs and concerns. In particular, integrating SSH and STEM expertise has provided an enhanced set of tools and processes for knowledge transfer, learning, and co-creation at the local level. In fact, our projects have offered a set of strategies and concrete measures to encourage dialogue and meaningful engagement. We find that participatory mechanisms should be developed for iterative stakeholder consultation procedures, which will foster a deeper understanding of the local context and address the concerns of vulnerable groups. Regional workshops have enabled project partners to identify central issues, capture the visions and ideas of citizens for a just transition, communicate research and modelling results back to various stakeholder groups, and ultimately highlight topics of interest that must be considered when adapting local plans and developing local projects.

Our project results also emphasise the importance of research in CCIRs and the need for continuous engagement of SSH and STEM scientists in CCRIs in order to identify the challenges but also empower stakeholders and citizens to have a say in the decisions that affect their livelihoods, facilitating processes that have their needs and concerns addressed.

Conclusions and Recommendations

The overall findings of the previous section led to the synthesis of policy recommendations towards people-centred just transition processes. First, involving local stakeholders with an intersectional approach is crucial in analysing the compensatory measures for holistically mitigating the negative impacts of decarbonisation policies or interventions. Defining new ways to meaningfully engage affected communities and citizens in the ongoing transition processes is essential. We call for empowering citizens to participate actively in the energy transition, e.g., through community-owned renewable energy projects, to gain more control over their power generation and consumption habits and increase their knowledge. Therefore, their opinion of what actually is a just transition becomes evidence-based.

Second, it is important to equip key actors with the transformative capacities to support the development of regional visions, plans, and narratives at the local level. Citizens should be informed and prepared to participate in the design and implementation of decarbonisation plans. The development of industrial and service sectors will depend on the entrepreneurial capacities and availability of skilled labour. Education and training activities are very important for improving human capital and regional development. More investment in research and innovation is required to meet the new challenges posed by just transition processes.

Third, the governance of just transitions requires a bottom-up approach and the decentralisation of power. To achieve this, structures and participatory mechanisms to encourage the active participation of a large number of stakeholders at the regional and local levels should be developed to ensure a diverse and inclusive social dialogue. Instead, the local stakeholders should be involved in the intersectional analysis of the compensatory measures for holistically mitigating the negative impacts of the policy or the intervention. To ensure compensatory measures are accessible to all sections of society that are either directly or indirectly affected by the transition policies or initiatives, it is imperative to identify measures that can enhance livelihoods before the implementation of the policy or initiative and that all local stakeholders, including citizens from all sections of society, are involved in that process.

Fourth, insights from science and practice stemming from participatory processes and methods and tools developed in different contexts to enable just transitions should be used in policymaking. The decarbonisation policies should not be politically motivated to please the voters of specific groups; instead, they should be based on scientific facts and a multiple-impact analysis of different measures. Governments should initiate multi-level governance and multi-stakeholder processes to provide agency to relevant local actors and enable a co-creation of solutions between science, policy, and practice.

Overall, we conclude that the active participation of well-equipped local stakeholders and citizens, the provision of appropriate tools and mechanisms for citizen engagement, science-based and co-creative policymaking, and the involvement of local stakeholders in the intersectional analysis of the compensatory measures planned for the vulnerable groups can accelerate the pace of just transition in CCIRs.

References

Abram, S., Atkins, E., Dietzel, A., Jenkins, K., Kiamba, L., Kirshner, J., Kreienkamp, J., Parkhill, K., Pegram, T., & Santos Ayllón, L. M. (2022). Just Transition: A whole-systems approach to decarbonisation. *Climate Policy, 22*(8), 1033–1049.

Akgüç, M., Arabadjieva, K., & Galgóczi, B. (2022). *Why the EU's patchy 'just transition' framework is not up to meeting its climate ambitions.* ETUI, The European Trade Union Institute. Retrieved 08:22, April 08, 2024, from https://www.etui.org/publications/why-eus-patchy-just-transition-framework-not-meeting-its-climate-ambitions

Carayannis, E. G., & Campbell, D. F. (2010). Triple Helix, Quadruple Helix and Quintuple Helix and how do knowledge, innovation and the environment relate to each other? A proposed framework for a trans-disciplinary analysis of sustainable development and social ecology. *International Journal of Social Ecology and Sustainable Development, 1*(1), 41–69.

Devine-Wright, P. (2007). Energy citizenship: Psychological aspects of evolution in sustainable energy technologies. In J. Murphy (Ed.), *Governing technology for sustainability* (pp. 63–88). Earthscan.

European Commission. (2019). *The European Green Deal COM (2019)* Publications Office of the European Union: Luxembourg; Publications Office of the European Union: Brussels, Belgium. https://eur-lex.europa.eu/resource.html?uri=cellar:b828d165-1c22-11ea-8c1f-01aa75ed71a1.0002.02/DOC_1&format=PDF

Garha, N. S., & Garcia Mira, R. (2023). *ENTRANCES deliverable 6.4. Knowledge co-production report.* https://entrancesproject.eu/project-deliverables/

Hanke, F., Guyet, R., & Feenstra, M. (2021). Do renewable energy communities deliver energy justice? Exploring insights from 71 European cases. *Energy Research and Social Science, 80*, 102244.

Hermwille, L., Schulze-Steinen, M., Brandemann, V., Roelfes, M., Vrontisi, Z., Keskülä, E., Anger-Kraavi, A., Trembaczowski, Ł, Mandrysz, W., Muster, R., & Zygmunt-Ziemianek, A. (2023). Of hopeful narratives and historical injustices—An analysis of just transition narratives in European coal regions. *Energy Research & Social Science, 104*, 103263.

JTM. (2021). *Just Transition Mechanism adopted by European Commission.* https://commission.europa.eu/strategy-and-policy/priorities-2019-2024/european-green-deal/finance-and-green-deal/just-transition-mechanism_en

KAPE, IEECP, BSERC, NTUA, CAC, AISVJ, FAEN, IRENA. (2023). *Knowledge sharing report.* Deliverable 3.1. JUSTEM project. https://ieecp.org/wp-content/uploads/2024/02/JUSTEM_D3.2-Knowledge-Sharing-Report.pdf

Koasidis, K., Nikolaev, A., Gaydarova, E., Karamaneas, A., Todorov, T., Georgiev, G., Irimie, S., Valmaseda, C., Gonzalez, I., Frankovic, A., Mazur, A., Ogrodniczuk, J., & Süsser, D. (2023). *Report on the current status of coal-dependent regions in the EU—JUSTEM "D2.1: Current Status of the Regions" and supplementary material.* Zenodo.

Lee, S., & Baumgartner, L. (2022). *How just transition can help deliver the Paris Agreement.* Report published by the United Nations Development Programme (UNDP). https://climatepromise.undp.org/sites/default/files/research_report_document/Just%20Transition%20Report%20Jan%2020.pdf

Michas, S., & Flamos, A. (2023). Are there preferable capacity combinations of renewables and storage? Exploratory quantifications along various technology deployment pathways. *Energy Policy, 174,* 113455.

Schot, J., Kanger, L., & Verbong, G. (2016). The roles of users in shaping transitions to new energy systems. *Nature Energy, 1,* 1–7.

Sovacool, B. K., David, J. H., Cantoni, R., Lee, D., Brisbois, M. C., Walnum, H. J., Dale, R. F., Rygg, B. J., Korsnes, M., Goswami, A., Kedia, S., & Goel, S. (2022). Conflicted transitions: Exploring the actors, tactics, and outcomes of social opposition against energy infrastructure. *Global Environmental Change, 73,* 102473.

Stavrakas, V., & Flamos, A. (2020). A modular high-resolution demand-side management model to quantify benefits of demand-flexibility in the residential sector. *Energy Conversion and Management, 205,* 112339.

Strasser, B. J., Baudry, J., Mahr, D., Sanchez, G., & Tancoigne, E. (2019). "Citizen science"? Rethinking science and public participation. *Science & Technology Studies, 32*(2), 52–76.

The European Commission. (n.d.). *Just Transition Platform.* Just Transition Platform. https://ec.europa.eu/newsroom/regio/newsletter-archives/45487

Wang, X., & Lo, K. (2021). Just transition: A conceptual review. *Energy Research & Social Science, 82,* 102291.

Open Access This chapter is licensed under the terms of the Creative Commons Attribution 4.0 International License (http://creativecommons.org/licenses/by/4.0/), which permits use, sharing, adaptation, distribution and reproduction in any medium or format, as long as you give appropriate credit to the original author(s) and the source, provide a link to the Creative Commons license and indicate if changes were made.

The images or other third party material in this chapter are included in the chapter's Creative Commons license, unless indicated otherwise in a credit line to the material. If material is not included in the chapter's Creative Commons license and your intended use is not permitted by statutory regulation or exceeds the permitted use, you will need to obtain permission directly from the copyright holder.

CHAPTER 9

Developing Equitable Maritime Spatial Planning in the EU: Case Studies from Portugal and Norway

Dina Abdel-Fattah, Misse Wester, Irene Martins, Sandra Ramos, and Stian K. Kleiven

Policy Highlights To achieve the recommendation stated in the title, we propose the following:

- Maritime spatial planning (MSP) is a mandated approach set forth by the EU to develop a common framework to manage maritime spaces.

D. Abdel-Fattah (✉) · S. K. Kleiven
UiT—The Arctic University of Norway, Harstad, Norway

D. Abdel-Fattah
Norwegian Meteorological Institute, Oslo, Norway

M. Wester
Lund University, Lund, Sweden

I. Martins · S. Ramos
University of Porto, Porto, Portugal

© The Author(s) 2024
E. Galende Sánchez et al. (eds.), *Strengthening European Climate Policy*, https://doi.org/10.1007/978-3-031-72055-0_9

- MSP helps to bring together the perspectives of diverse stakeholders, although challenges exist to ensure it equitably reflects their needs and concerns.
- The Portuguese case shows the importance of including top-down and bottom-up participation mechanisms to ensure all stakeholders are integrated into the process.
- MSP can be used to protect sensitive areas and safeguard marine life, but the Norwegian case shows that political buy-in is imperative for its successful implementation.
- Combining natural and social sciences in marine planning ensures the integration of diverse views and helps to develop a robust and equitable process.

Keywords Maritime Spatial Planning (MSP) · Equity · Fairness · Environmental impact · Stakeholder participation

Introduction

The EU has set an ambitious agenda to combat climate change, in which renewable energies play a fundamental role in achieving climate neutrality by 2050 (European Commission, 2019). In this scenario, offshore renewable energy sources, including marine energy, will have increasing importance in the European energy system. Therefore, Maritime Spatial Planning (MSP) is an important tool to ensure the sustainable deployment of the blue energy sector within the framework of the European Integrated Maritime Policy (García et al., 2020). In 2014, the EU adopted the Maritime Spatial Planning Directive to promote the sustainable development of marine areas and the responsible use of its resources, requiring each Member State to adopt and implement a Maritime Spatial Plan (as of 2023, all member states have either adopted or are in the process of adopting these plans). This Directive, in particular Articles 4–7, 10–12, sets forth seven minimum requirements, requiring maritime planning to (Friess & Grémaud-Colombier, 2021):

1. take into account land-sea interactions;
2. take into account environmental, economic, and social aspects, as well as safety aspects;
3. aim to promote coherence between maritime spatial planning and the resulting plan or plans and other processes, such as integrated coastal management or equivalent formal or informal practices;
4. ensure the involvement of stakeholders;
5. organise the use of the best available data;
6. ensure transboundary cooperation between Member States;
7. promote cooperation with third countries.

In addition, the United Nations Educational, Scientific and Cultural Organization's Intergovernmental Oceanographic Commission (UNESCO-IOC) developed in 2021 an International Guide on Marine/Maritime Spatial Planning, outlining the main and most important aspects to be included in marine spatial plans.

Despite the common understanding, acceptance, and adoption of MSP across the EU as a tool to jointly consider industrial, economic, and social objectives, limitations and challenges exist in its development and implementation as a decision-making tool and conflict-resolution resource. Although the ambition is to prioritise and optimise the use of maritime resources and areas, conflicts often arise when software-based decisions, such as those arising from MSP, are perceived by stakeholders as either complex or untrustworthy. Therefore, MSP-based decisions should be inclusive, socially fair, equitable, and flexible in order to be effective (Boussarie et al., 2023).

In this chapter, we review the design and implementation process of maritime spatial plans in two key case studies—Portugal and Norway—to highlight the critical discussions that have taken place, and are currently underway, regarding how to utilise MSP as a tool for the just and equitable distribution and use of maritime areas. The interdisciplinary author team, whose expertise spans social sciences, marine ecology, climatology, and risk analysis, reviewed the MSP processes and plans in both countries, while critically looking into the public discourses and current events regarding maritime development. An intensive writing retreat was conducted as part of this work, where the authors dived deeply into these topics, bringing up important cross-cutting interdisciplinary considerations discussed further below. Thanks to this interdisciplinary approach, we base our arguments not only on data provided by the natural sciences but also on a wider range of perspectives that include the issues of justice and fairness.

Experiences with MSP: Case Studies from Portugal and Norway

Portugal

The *Maritime Spatial Plan* for Portugal (acronym in Portuguese: PSOEM) was established in 2014 through the Basic Law of the National Maritime Spatial Planning and Management Policy targeting the temporal and spatial management of human activities operating in the Portuguese maritime area. PSOEM is an instrumental framework for planning the entire Portuguese maritime space, including inland maritime waters, territorial seas, exclusive economic zones, and continental shelves extending beyond 200 nautical miles. It is considered the *situation plan* as it delineates the spatial and temporal distribution of current and potential uses and activities and identifies relevant areas of conservation, including biodiversity and underwater cultural heritage values. It also includes critical networks and structures associated with national defence, internal security, and civil protection. Functioning as a pivotal tool for maritime policy, this plan aims to optimise compatibility among competing uses or activities, with the overarching goal of enhancing the economic utilisation of the marine environment and mitigating the adverse impacts of human activities on marine ecosystems.

Another type of national MSP instrument corresponds to the *allocation plans* (Calado et al., 2024) that are meant to identify—and allocate areas to—specific "new" uses that were not yet included in the PSOEM (either as existing or potential uses). Upon approval, these plans become automatically integrated into the PSOEM, and may be carried out either by public or private initiatives, although in this case, there must always be a public entity responsible for the plan. Hence, PSOEM and the allocation plans serve as the enabling mechanisms for the issuance of a *Permit for Private Use of the National Maritime Space (PPUNMS)*, which is the specific instrument to manage the private use of Portuguese maritime space.

PPUNMS are formalised through the issuance of a *Title of Private Use of the Maritime Space (TUPEM)* that has a specific duration and can be of three types: concession, licence, or authorisation, irrespective of the nature and legal structure of the recipient. Several TUPEM modalities are foreseen, namely in aquaculture; energy resources, including exploration of renewable energies, gas, and oil; research; recreation, sports, and tourism; immersion of waste/dredging; infrastructure and equipment; as well as other uses or activities of industrial nature.

Notwithstanding, some gaps have been recognised in PSOEM, particularly regarding the complexity and state of marine ecosystems and the impact of certain existing and potential activities in the marine environment. For example, for some emerging activities such as marine biotechnology, metallic mineral resources, and geological carbon storage, the situation plan only characterises the activities without presenting potential areas or guidelines for compatibility between activities and minimising impacts on the environment (Calado et al., 2024).

Another point of concern relates to areas with protection status, including the Natura 2000 Network and marine protected areas, where special care must be ensured to comply with their management guidelines while adopting complementary measures to minimise possible negative impacts. In addition, the participation of stakeholders in the Portuguese MSP has been mostly top-down, from the initial definition of objectives to the final stages, despite being recommended that stakeholders be integrated throughout the entire process (Gómez-Ballesteros et al., 2021).

To account for some of the current limitations of PSOEM, a *Strategic Environmental Assessment* (SEA) process was implemented, aiming at optimised environmental integration by assessing opportunities and risks of actions, evaluating, and comparing alternative development scenarios before decision-making. Adding up to this, specific methodologies have been developed to address missing links between different instruments (e.g., EU MSP Directive, EU Marine Strategy Framework Directive, EU Strategic Environmental Assessment Directive) (Calado et al., 2021).

Besides the existence of a legal framework for MSP in Portugal, the PSOEM made possible the existence of a geoportal with relevant information in a digital, open-access format that is continuously updated, which facilitates transparency, information sharing, and consultation by all interested parts, from authorities to investors, and thus supports and expedites decision-making (Calado et al., 2024). Overall, the Portuguese experience is positive, although further improvements should be considered given the complexity and dynamic nature of the process.

Norway

Although not an EU member state, Norway is part of the European Economic Area (EEA). However, the EEA follows a smaller thematic and geographical area of operation than the EU. For example, directives related to conservation, agriculture, fisheries, and EU habitats are not included as part of the EEA agreement's scope. In addition, the EEA agreement only applies to the Norwegian territory, which is the area between the main border and up to 12 nautical miles, while the rest of the regions follow constitutional law and international agreements (Schütz & Johansen, 2023).

The MSP Directive is not included in the EEA agreement, and therefore Norway is not required to abide by it. Still, Norway is considered one of the European countries with an established system to manage their seas, and in many ways, the country's legal framework already included the requirements of the EU MSP Directive. Norway's strategy for ecosystem-based management includes the Barents Sea, the Norwegian Sea, the North Sea, and Skagerrak. Also, maritime spatial use follows different laws. For instance, in coastal regions, it follows the National Planning and Building Act, the Land and Water Act, the Marine Resources Act, and the Aquaculture Act. In addition, maritime spatial use is decided by municipal spatial plans (Schütz & Johansen, 2023).

In 2022, the Norwegian Government developed a roadmap with principles for maritime spatial use. This roadmap is meant to create predictability and a foundation for the coexistence of different maritime industries, such as, but not limited to, wind energy, carbon capture and storage, green shipping, sustainable seafood production, kelp production, and carbon binding. Every fourth year, a message about administration plans for the spatial sea areas is made; the next one being developed in 2024 with a focus on sustainable seas.

One of the most controversial decisions of this process was taken in June 2023, when the Norwegian Government proposed deep-sea mining in a 280,000 km^2 region in the Arctic between Svalbard, Greenland, Iceland, and Jan Mayen. In January 2024, the Norwegian Parliament approved this proposition despite clear advice and rejection from the Norwegian Environment Agency and the Norwegian Institute of Marine Research, due to the existing knowledge gaps about its impact and consequences for various species, the environment, and fisheries (Stortinget, 2023).

Therefore, different companies can apply to start deep-sea mining in the proposed region to extract metals such as copper, cobalt, gold, and other rare minerals that can be found in manganese crusts (Stortinget, 2023). This has led to debates and dissatisfaction among numerous international and national environmental organisations, while other countries, US states, and even companies have set in place or demanded a moratorium/ban on deep-sea mining (WWF, 2024). Put simply, local and global actors are worried—based on the current scientific understanding—about the unknown social and environmental consequences; while some industrial actors, such as the companies Loke Marine Minerals and Green Minerals, argue that more minerals are needed for the technological development enabling the green transition.

One of the main concerns is the impact of deep-sea mining on some organisms endemic to this region, such as sponge grounds that grow on manganese crusts. Sponge grounds are areas where a high number of sponges live and can stretch for several kilometres. These sponge aggregations are considered threatened by the Convention for the Protection of the Marine Environment of the North-East Atlantic (OSPAR Convention) and are classified as vulnerable marine ecosystems. Also, other species are dependent on these sponge grounds, as they provide structural habitats.

Overall, MSP is an important process to help facilitate difficult but important conversations. By listening to research communities, organisations, and other stakeholders, the hope is that maritime areas can be utilised in the best possible manner, taking into consideration both environmental and social impacts. However, the recent approval of deep-sea mineral mining shows that MSP is not always followed entirely. This case study shows that, although MSP can be used to safeguard marine life, several challenges remain. The exploitation of protected areas for commercial uses or green transitions risks damaging both environmentally sensitive areas and disrupting trust and confidence between the different actors in maritime planning. Lessons learned from deliberation around other topics should be included in future work about maritime planning, particularly on deep-sea mining.

Conclusions and Recommendations

As shown in the Portugal and Norway case studies, there are tensions between different stakeholders relating to the governance of maritime issues. Indeed, since the development of MSP, critique has been raised that this process ignores existing power relations and rather favours scientific knowledge at the expense of other perspectives (Flannery et al., 2018). For example, the introduction of commercial activities such as aquaculture or offshore wind farms can have local implications that are not captured in risk assessments based on quantitative scientific calculations. For instance, these calculations do not take into account the changing physical environment on local social identities or the changes in housing prices. Also, if traditional livelihoods are threatened, even a small risk can be seen as unacceptable by local communities. Indeed, MSP can be seen as a process to increase "blue production" rather than promote sustainable blue growth as intended by the EU.

One explicit aim of this tool is to engage with all stakeholders, but research indicates that previous experiences have much room for improvement. The planning process usually involves making decisions along several phases. However, it is usually not until the process reaches the operational stage—where a final decision has to be made—that stakeholders are involved (Flannery et al., 2018). Involving stakeholders late in the process is often a mere act of tokenism, as they can only react to the proposal in front of them, and often leads to increased conflict arising from power inequalities.

If MSP intends to address all interests and concerns related to maritime-both nearshore and offshore-management in a fair and equitable manner all stakeholders, be they large companies or local communities, need to be able to contribute to the whole planning process and need to be provided with the same opportunities and responsibilities. This also includes fairness in access to resources and information, transparency about how participation is linked to decision-making, and willingness to learn (Chilvers, 2008). In addition, the process can significantly benefit from the facilitation by an independent agent.

Based on the two case studies discussed above, as well as the MSP literature, we conclude that MSP can be an effective platform to bring stakeholders together to make critical, yet difficult, decisions regarding maritime spatial use. Therefore, our main policy recommendation is that

EU strategies for Maritime Spatial Planning must ensure the fair and equitable representation of all stakeholders' needs and concerns to balance environmental, social, and economic goals.

The involvement of all stakeholders, allowing for a multiplicity of perspectives to be included, is crucial for reaching democratic and equitable decisions. This means that representation based on numbers, e.g., one representative per organisation, can lead to the prevalence of existing power dynamics. Instead, representation needs to be equitable among sectors, where industrial interests are given the same power of influence as environmental or cultural interests. When interests compete, there are bound to be situations where not all stakeholders will be satisfied with the final solution. For example, increasing mining for the green transition might not be compatible with preserving biodiversity. In this context, stakeholders with large resources, such as private companies, will be usually more likely to assert their positions than local environmental groups arguing for conservation. To avoid this, there is a need to create fair and transparent processes, where stakeholders are active participants in reaching a joint decision, rather than passive actors asked to justify an already-made decision (Chilvers, 2008). If the process is perceived as fair, chances are that most stakeholders will leave the process with an increased understanding of other stakeholders' perspectives and needs, even if disagreement on the final decision is difficult to avoid.

To conclude, we have highlighted in this chapter how maritime spatial planning faces the same challenges as many models and frameworks that seek to engage stakeholders in the decision-making processes affecting our collective future. In developing this contribution, the interdisciplinary collaboration between natural and social scientists proved to be an enriching experience, resulting in fruitful discussions among the authors that stressed the need to consider equity aspects. Maritime spatial planning is a fairly new tool, with an opportunity to help, rather than hinder, equity. We, therefore, urge decision-makers to learn from previous experiences and utilise this tool to promote fair and equitable stakeholder involvement and arrive at decisions that can be respected by everyone involved. In this way, MSP has the potential to build trust and pave the way towards a truly sustainable governance of the marine environment.

References

Boussarie, G., Kopp, D., Lavialle, G., Mouchet, M., & Morfin, M. (2023). Marine spatial planning to solve increasing conflicts at sea: A framework for prioritizing offshore windfarms and marine protected areas. *Journal of Environmental Management, 339*, 117857.

Calado, H., Gutierrez, D., Pegorelli, C., Kirkfeldt, T.S., Hipólito, C., Moniz, F., McClintock, W., Vergílio, M., Guerreiro, J., & Papaioannou, E. (2021). A tailored method for strategic environmental assessment in maritime spatial planning, *Journal of Environmental Assessment Policy and Management, 23*.

Calado, H., Frazão Santos, C., Quintela, A., Fonseca, C., & Gutierrez, D. (2024). The ups and downs of maritime spatial planning in Portugal. *Marine Policy, 160*, 105964.

Chilvers, J. (2008). Deliberating competence: Theoretical and practitioner perspectives on effective participatory appraisal practice. *Science, Technology, & Human Values, 33*(3), 421–451.

European Commission. (2019). *The European Green Deal Communication*. https://commission.europa.eu/publications/communication-european-green-deal_en

Flannery, W., Healy, N., & Luna, M. (2018). Exclusion and non-participation in marine spatial planning. *Marine Policy, 88*, 32–40.

Friess, B., & Grémaud-Colombier, M. (2021). Policy outlook: Recent evolutions of maritime spatial planning in the European Union. *Marine Policy, 132*, 103428.

García, P. Q., Ruiz, J. A. C., & Sanabria, J. G. (2020). Blue energy and marine spatial planning in Southern Europe. *Energy Policy, 140*, 111421.

Gómez-Ballesteros, M., Cervera-Núñez, C., Campillos-Llanos, M., Quintela, A., Sousa, L., Marques, M., Alves, F. L., Murciano, C., Alloncle, N., Sala, P., & Lloret, A. (2021). Transboundary cooperation and mechanisms for maritime spatial planning implementation. SIMNORAT project. *Marine Policy, 127*, 104424.

Schütz, S. E., & Johansen, E. (2023). *Faglig grunnlag for overordnede prinsipper for arealbruk til havs*. Retrieved 31 January 2024 from https://bora.uib.no/bora-xmlui/handle/11250/3062756

Stortinget. (2023). *Mineralverksemd på norsk kontinentalsokkel - opning av areal og strategi for forvaltning av ressursane*. Retrieved 31 January 2024 from https://www.stortinget.no/nn/Saker-og-publikasjonar/Saker/Sak/?p=94807

WWF. (2024). *Global moratorium on deep seabed mining*. Retrieved 6 May 2024 from https://www.stopdeepseabedmining.org/statement/

Open Access This chapter is licensed under the terms of the Creative Commons Attribution 4.0 International License (http://creativecommons.org/licenses/by/4.0/), which permits use, sharing, adaptation, distribution and reproduction in any medium or format, as long as you give appropriate credit to the original author(s) and the source, provide a link to the Creative Commons license and indicate if changes were made.

The images or other third party material in this chapter are included in the chapter's Creative Commons license, unless indicated otherwise in a credit line to the material. If material is not included in the chapter's Creative Commons license and your intended use is not permitted by statutory regulation or exceeds the permitted use, you will need to obtain permission directly from the copyright holder.

CHAPTER 10

Bringing in Ethics: A Multi-stakeholder Approach to Manage the Transition to Low-Carbon Construction

Michal Plaček, *Vladislav Valentinov, Roman Fojtík, František Ochrana, and Martina Peřinková*

Policy Highlights To achieve the recommendation stated in the title, we propose the following:

- Policymakers should organise fair and balanced stakeholder engagement processes to address the key ethical trade-offs of the transition to low-carbon construction.
- The public sector should take a proactive role in leading the low-carbon transition and build strong cross-sectoral partnerships.

M. Plaček (✉) · F. Ochrana
Faculty of Social Sciences, Charles University, Prague, Czech Republic
e-mail: 62666039@fsv.cuni.cz

F. Ochrana
e-mail: frantisek.ochrana@fsv.cuni

V. Valentinov
The Leibniz Institute of Agricultural Development in Transition Economies (IAMO), Halle (Saale), Germany
e-mail: Valentinov@iamo.de

© The Author(s) 2024
E. Galende Sánchez et al. (eds.), *Strengthening European Climate Policy*, https://doi.org/10.1007/978-3-031-72055-0_10

- Public sector leadership demands transparency and accountability which are crucial for navigating ethical trade-offs with the various stakeholders in the construction sector.
- Policymakers should build a culture of sustainability and cultivate a shared understanding of ethics and values among all stakeholders.
- To find creative solutions to the complex challenges associated with the transition to low-carbon construction, SSH and STEM collaboration should be supported.

Keywords Low-carbon construction · Ethics · Stakeholder Participation · Transparency · Public sector leadership

Introduction

This chapter examines the ethical challenges of transitioning to low-carbon construction and offers policy recommendations at the EU level to address them. Currently, the construction sector globally contributes to 37% of emissions and employs approximately 7% of the workforce (World Green Building Council 2023). Transforming this sector is critical for achieving the UN Sustainable Development Goals (e.g., SDGs 1–2, SDGs 11–12, and SDG17), as it has great potential to reduce resource and energy consumption, increase the use of renewable energy, minimise environmental degradation and waste, and enhance occupant health and comfort (González-Díaz & Garcia-Navarro, 2011, p. 295).

We argue that the transition to low-carbon construction is not solely a technological issue but involves complex decision-making processes, including weighing short- and long-term costs and benefits across society and stakeholders. This process is inherently ethical. Ethical concerns in the

---

R. Fojtík · M. Peřinková
VŠB Technická Univerzita, Ostrava, Czech Republic
e-mail: fojtikr@fld.czu.cz

M. Peřinková
e-mail: martina.perinkova@vsb.cz

construction sector include conflicts of interest, financial integrity, corruption, consumer and employee privacy, and ethical advertising (Sohail & Cavill, 2008). Furthermore, research suggests that the professional ethics of the civil engineering sector may conflict with sustainability goals while also acknowledging the above-mentioned ethical issues within the construction sector (Mares-Nasarre et al., 2023).

The ethical complexity of transitioning to low-carbon construction requires an interdisciplinary approach. In this chapter, we, therefore, integrate insights from Public Policy, Sociology, Ethics, Materials Engineering, and Architecture to address how to manage ethical transitions. To achieve this, we utilised a mixed-method approach combining a systematic literature review and a workshop.

The first step was a systematic literature review aimed at understanding the current state of knowledge on the ethical challenges of the low-carbon transition process. The literature review, conducted in September 2023 using the Web of Science database, identified 234 records using keywords such as "low carbon construction + ethics" and "sustainable construction + ethics". The review affirms that the ethical complexity inherent in the transition to low-carbon construction spans various scientific sub-disciplines, resulting in significant knowledge fragmentation. The majority of articles were classified under environmental sciences (33 articles), environmental studies (33 articles), ethics (27 articles), green & sustainable science & technology (26 articles), business (18 articles), and management (18 articles), among others. After an abstract review, 35 articles were selected for in-depth examination. The systematic literature review also informed the development of a scenario to be used at the collaborative workshop, which formed the second step of our mixed-method approach.

The workshop facilitated collaboration among researchers from various fields and aimed to provide a space for the participants to collaborate on new technologies, services, products, or systems (Duchková, 2023). Participant recruitment involved purposive sampling to ensure the representation of various stakeholders. Nine respondents were selected, including an applied ethicist, materials engineer, construction company owner, bridge designer, ministry employee, sustainability researcher, university transfer representatives, and a former senior public sector employment manager. Given the diverse backgrounds of the workshop participants (education and experience), we provided explanations and definitions of basic concepts, such as ethics in low-carbon construction.

At the workshop, conducted online in December 2023, participants received the above-mentioned scenario outlining key points, definitions, and questions derived from the systematic literature review. Participants were then asked to collaboratively define key concepts, unify knowledge levels, and discuss creative solutions to the ethical challenges of the transition to low-carbon construction. Moderated by one of the study authors, the workshop lasted approximately two hours and was recorded and transcribed. The data were coded using deductive and inductive approaches to explore new phenomena and mechanisms not previously described within the research framework. To minimise bias, the analysis was conducted by a team member different from the moderator (Jelínková et al., 2023). Participants were provided with a draft of the results for independent comments after the workshop.

Results

Ethical Dimensions and Stakeholders in Transitioning to Low-Carbon Construction

Our examination has revealed that the ethical aspects of transitioning to low-carbon construction are fundamentally grounded in the human awareness of belonging to a larger interconnected system and the responsibility for life on this planet (Grunwald, 2001). However, effectively addressing this complexity requires breaking down these holistic insights into the interests of multiple stakeholders, which can often conflict (Rostamnezhad & Thaheem, 2022).

During the literature review, we identified various stakeholders of relevance for the low-carbon transition in the construction sector, including government entities, developers, architects, engineers, and others. Participants in the workshop emphasised the importance of also including independent certification and compliance authorities, universities, and information brokers in the overview of relevant stakeholders.

Participants were also asked to identify the stakeholders they considered "at-risk" due to negative effects of transition processes, such as increased costs, financial loss, or disruptions in the supply chain. Surprisingly, construction firms were not perceived as "at-risk stakeholders" during the transition; instead, materials producers were considered the most vulnerable. Discussions also touched upon the risk of unemployment. However, the participants noted even greater challenges regarding demographic issues and problems of securing the available workforce due to an ageing population.

Our workshop revealed numerous potential ethical trade-offs faced by the diverse range of stakeholders (see Table 10.1).

These diverse stakeholders and the various trade-offs they face highlight the intricate web of ethical considerations inherent in the transition to low-carbon construction, underscoring the need for comprehensive and inclusive approaches to address them.

Resolving Trade-Offs Through Fair Stakeholder Interaction

Addressing these trade-offs effectively ultimately requires fair and balanced stakeholder interaction processes (Valentinov, 2023). Our systematic literature review highlights the significance of cross-sectoral partnerships in stakeholder interaction (Andrews & Entwistle, 2010). These partnerships involve collaboration between the public sector, business, and the non-profit sector, aiming to address society's main challenges.

Building on these findings, our workshop results emphasise the crucial role of clear leadership. Participants in the living lab expressed their belief in the efficiency of markets during a low-carbon transition but stressed the vital role of the public sector in providing leadership, directing the transition, and aligning diverse stakeholder interests. However, they also acknowledged the need for the public sector to enhance its ability to lead change and collaborate with different stakeholder groups. "*Creating a culture of sustainability, defined by values, beliefs, and behaviour favouring sustainability*" (Yip Robin & Poon, 2009, p. 3617), was deemed essential.

Another workshop outcome highlighted the importance of ethical precautions in designing financial support instruments such as grants, loans, and procurement. Public procurement, given its significant GDP allocation, was considered a key instrument. Participants stressed the necessity for transparent, value-for-money resource allocation to maintain legitimacy and credibility. Intelligent investment, which considers life-cycle costs and societal effects beyond merely pursuing the lowest price, was also emphasised. These tools should be institutionalised as long-term policy instruments.

The workshop participants also emphasised information and transparency as integral components of accountability. They highlighted the

Table 10.1 Overview of key stakeholders and ethical trade-offs

Stakeholder type	Actors	Trade-Offs	Considerations
Government	• Central government • Regional government • Local government	• Leadership • Capacity allocation • Environmental stewardship • Cross-sectoral collaboration	• Acting as a leader or allowing others to behave according to their rationality • Allocating capacity to the low-carbon transition versus other government tasks • Balancing compliance-oriented behaviour with accountability pressures, and potential weakening of discipline for specific stakeholder involvement
Companies	• Main contractors • Subcontractors • Suppliers • Managers • Employers • Developers	• Profit vs. environmental and societal goals	• Balancing long-term and short-term effects • Managing interests within the supply chain • Navigating between conservative biases and fostering innovation

(continued)

Table 10.1 (continued)

Stakeholder type	Actors	Trade-Offs	Considerations
Professional communities	• Architects • Structural engineers • Electrical and mechanical engineers • Surveyors	• Professional ethics vs. personal interests	• Balancing transparency and openness of communication versus public comprehensibility • Balancing precaution versus speed of innovation
Environmental lobby groups	• NGOs • Lobbyists • Think tanks	• Accountability vs. members' interests	• Balancing socio-environmental impact and lobby group interests
Media organisations	• Newspapers • TVs & radios • New media • Journalists • Professional associations • Owners • Shareholders	• Shareholder interests vs. the interests of readers	• Balancing revenue from advertisements versus firms' environmental impacts • Reducing the complexity of information and making it clear to the reader • Balancing the verification of the veracity of the information with the speed of publication
Environmental management accreditation bodies and independent certification organisations	• Public sector organisations • Private sector organisations • Consultancies	• Ethical values and integrity vs. the interests of clients	• Focusing on accountability, transparency, and fulfilling professional standards

(continued)

Table 10.1 (continued)

Stakeholder type	Actors	Trade-Offs	Considerations
Higher education institutions, research institutions, and training providers	• Public and private universities • Public and private research institutions • Education companies	• Quality and integrity, speed of publication, independence of research vs interests of funders	• Scientific results should be freely available to all, and the underlying data should be accessible and meet the requirements of transparency and replicability • Communicate scientific results inclusive for all social groups • Mechanisms should be put in place to protect researchers with critical views on certain topics from institutional pressure
Funding institutions	• Public institutions • Private institutions	• Pressure to deliver results versus uncertainty in scientific knowledge	• Openness in funding and the originality and disruptiveness of projects should be evaluated, not previous results, seniority, and perceived prestige • Accept the risk of failure as a normal part of scientific work
General public		• Personal bias vs innovation, rationality vs emotions	• Balance environmental engagement with privacy considerations

Source Authors

importance of Environmental, Social, and Governance (ESG) reporting to encourage actors to prioritise sustainability. Information asymmetry was identified as a key issue, with suggestions that universities and certification authorities play a crucial role in presenting scientific study results efficiently to reduce transaction costs while maintaining credibility. The discussion also touched upon codes of ethics and corporate social responsibility documents, with participants noting a need for specific commitments rather than generalities.

In navigating ethical challenges successfully, the workshop participants emphasised the importance of ensuring that all levels of government possess sufficient capacity, including administration and resources, for policy implementation. Local governments, in particular, faced potential challenges due to complicated legislation and insufficient support for local politicians. Emphasising value for money was deemed vital for demonstrating the meaningful use of taxpayer money and legitimising implementation.

Implementing ethical solutions was viewed as a long-term commitment that required focus on the entire supply chain. Each part of the supply chain might encounter specific challenges of varying intensities. Therefore, a comprehensive approach addressing these challenges at each level is essential for successful and ethical policy implementation.

Conclusions and Recommendations

This chapter aims to identify the main ethical challenges of transitioning to low-carbon construction and to present recommendations for EU-level policymakers. These recommendations, collectively addressing the multifaceted nature of the transition, aim to guide policymakers in developing effective strategies that consider the technical, social, political, and economic aspects of the low-carbon construction shift. The advantage of these recommendations is that they can be scaled up to lower levels of government, such as national and local governments.

On the basis of our analysis, four key ethical principles can be deduced for EU policymakers in the transition to low-carbon construction:

1. **Fair and Balanced Stakeholder Engagement**

   Recognise numerous stakeholders: Policymakers should acknowledge and engage the diverse range of relevant stakeholders, including actors such as developers and architects, universities, certification bodies, and the public.

   Address trade-offs: In these stakeholder engagement processes the identified trade-offs must be addressed and policymakers should strive to create solutions that fairly balance the competing interests of each stakeholder group.

2. **Leadership and Collaboration**

   Proactive public sector: Governments at all levels should play a proactive role in leading the low-carbon transition, facilitating collaboration between various stakeholders. This leadership should be ethical, transparent, and focused on the common good. This entails establishing ambitious goals, developing clear roadmaps, and providing incentives for low-carbon practices.

   Cross-sectoral partnerships: Building strong partnerships among the public sector, private businesses, and non-profit organisations is crucial for addressing complex challenges like the low-carbon transition. Partnerships involve creating platforms for diverse stakeholders—construction companies, architects, universities, etc.—to collaborate effectively, share knowledge, and devise solutions.

3. **Transparency and Accountability**

   Ethical Policy Design: Policies and regulations concerning low-carbon construction should prioritise environmental benefits, ensure fair competition, and mitigate potential harm to vulnerable stakeholders like construction workers.

   Transparency and Accountability: Public sector leadership demands open communication about goals, progress, and challenges. Transparency in decision-making processes and accountability for policy outcomes foster trust with stakeholders and promote responsible implementation.

   Clear and accessible information: Policymakers should ensure transparent communication of information about the transition, its goals, and potential challenges. Complex scientific data should be presented clearly and accessibly, potentially with the help of universities and certification bodies.

ESG reporting: Policymakers should encourage actors across the construction industry to adopt Environmental, Social, and Governance (ESG) reporting practices to foster accountability and transparency.

Ethical financial instruments: Policymakers should design financial support instruments, such as grants and loans, with ethical considerations in mind, including factors like life-cycle cost analysis and responsible resource allocation.

4. Building a Culture of Sustainability

Shared values: A shared understanding of sustainability values must be cultivated and ethical behaviour encouraged across all stakeholders through education, awareness campaigns, and promoting a culture of responsibility.

Long-term commitment: Policymakers should create sustainable policy frameworks that consider the entire supply chain and address the diverse challenges faced by different stakeholders throughout the process.

Adhering to these four key ethical principles will enable EU policymakers to foster a more inclusive, equitable, and sustainable approach to the low-carbon construction transition, benefiting all stakeholders and the planet. These principles must be integrated into everyday public leadership practices, serving as fundamental tools for successful transition.

Research on the implementation of sustainability policies shows that one of the key success factors is the institutionalisation of policies not only at the system level but also at the level of individual organisations. The following tools are recommended for successful institutionalisation:

- **Long-term policy frameworks**: Develop comprehensive policies and regulations that transcend short-term political cycles. These frameworks should establish clear industry expectations, provide long-term stability, and encourage investment in low-carbon solutions.
- **Capacity building**: Invest in enhancing the government's internal capacity to manage the transition effectively. This includes training

staff on sustainability principles, life-cycle costing, and stakeholder engagement.
- **Financial instruments with ethics**: Design financial tools such as grants, loans, and tax breaks to incentivise ethical low-carbon practices. Consider factors like life-cycle costs, local sourcing of materials, and fair labour practices.
- **Public procurement**: Use government purchasing power to promote low-carbon construction. Establish clear sustainability criteria for public construction projects and prioritise companies with strong Environmental, Social, and Governance (ESG) practise.
- **Knowledge sharing and collaboration**: Facilitate knowledge sharing among research institutions, universities, and the construction industry. Foster collaboration in developing and implementing innovative low-carbon construction technologies.
- **Independent oversight and enforcement**: Establish independent oversight bodies to monitor and enforce regulations, ensuring ethical practices throughout the construction supply chain.

By incorporating these ethical principles and utilising appropriate tools, EU policymakers can successfully contribute to the transition to low-carbon construction, fostering a sustainable future for all.

The main defining feature of the transition to low-carbon construction is complexity. This problem cannot be reduced to a technical or ethical dimension alone; collaboration between STEM and SSH can be a tool to address these challenges. We can expect that in the future, the boundaries between STEM and SSH will increasingly merge.

References

Andrews, R., & Entwistle, T. (2010). Does a cross-sectoral partnership deliver? Empirical exploration of public service effectiveness, efficiency, and equity. *Journal of Public Administration Research and Theory, 20*(3), 679–701.

Duchková, H. (2023). *Living Labs*. Engagement methods for climate, energy and mobility transitions. No. 6. Cambridge: SSHCentre.

González-Díaz, M. J., & García-Navarro, J. (2011). *Anthropocentric and non-anthropocentric values as the basis of the new sustainable paradigm in architecture*.

Grunwald, A. (2001). Application of ethics to engineering and the engineer's moral responsibility: Perspectives for a research agenda. *Science and Engineering Ethics, 7*(3), 415–428.
Jelínková, M., Plaček, M., & Ochrana, F. (2023). The arrival of Ukrainian refugees as an opportunity to advance migrant integration policy. *Policy Studies*, 1–6.
Mares-Nasarre, P., Martínez-Ibáñez, V., & Sanz-Benlloch, A. (2023). Analyzing the sustainability awareness and professional ethics of civil engineering bachelor's degree students. *Sustainability, 15*(7), 6263.
Rostamnezhad, M., & Thaheem, M. J. (2022). Social sustainability in construction projects—A systematic review of assessment indicators and taxonomy. *Sustainability, 14*(9), 5279.
Sohail, M., & Cavill, S. (2008). Accountability to prevent corruption in construction projects. *Journal of Construction Engineering and Management, 134*(9), 729–738.
Valentinov, V. (2023). Sustainability and stakeholder theory: A processual perspective. *Kybernetes, 52*(13), 61–77.
Yip Robin, C. P., & Poon, C. S. (2009). Cultural shift towards sustainability in the construction industry of Hong Kong. *Journal of Environmental Management, 90*(11), 3616–3628.

Open Access This chapter is licensed under the terms of the Creative Commons Attribution 4.0 International License (http://creativecommons.org/licenses/by/4.0/), which permits use, sharing, adaptation, distribution and reproduction in any medium or format, as long as you give appropriate credit to the original author(s) and the source, provide a link to the Creative Commons license and indicate if changes were made.

The images or other third party material in this chapter are included in the chapter's Creative Commons license, unless indicated otherwise in a credit line to the material. If material is not included in the chapter's Creative Commons license and your intended use is not permitted by statutory regulation or exceeds the permitted use, you will need to obtain permission directly from the copyright holder.

CHAPTER 11

Integrating Multispecies Justice Approach for Climate Risk Management in Forest Areas of Mediterranean Europe

Ethemcan Turhan, Cem İskender Aydın, Nurbahar Usta Baykal, and İsmail Bekar

Policy Highlights To achieve the recommendation stated in the title, we propose the following:

- Euro-Mediterranean forests face specific climate risks topped with demographic, economic, and societal pressures which call for a different approach to climate risk management.
- Interdisciplinary approaches in fire ecology, conservation biology, ecological economics, and political ecology unveil the emotional connection between humans and non-humans.

E. Turhan (✉)
Department of Spatial Planning and Environment, Groningen, The Netherlands
e-mail: e.turhan@rug.nl

C. İ. Aydın
Institute of Environmental Sciences, Boğaziçi University Hisar Campus, Istanbul, Turkey
e-mail: cem.aydin@bogazici.edu.tr

© The Editor(s) (if applicable) and The Author(s) 2024
E. Galende Sánchez et al. (eds.), *Strengthening European Climate Policy*,
https://doi.org/10.1007/978-3-031-72055-0_11

- Instead of end-of-pipe fire suppression, policy attention should focus on fuel build-up in the landscape and centre on the "state of shared fragility" between humans and more-than-humans.
- To prevent the firefighting trap, fire management and adaptation policies should be reviewed to incorporate scientific expertise, local ecological knowledge, and traditional practices of forest communities.
- Locally grounded, value-based responses such as IPBES's Nature's Contribution to People (NCP) framework hold promising potential for multispecies justice in Euro-Mediterranean forests.

Keywords Euro-Mediterranean forests · Wildfire impacts · Climate risk management · Multispecies justice · Value-based approaches

Introduction

The IPCC's 6th Assessment Report highlights the urgency of the climate crisis, predicting a potential 96–187% increase in burnt forest areas under a 3 °C temperature rise by the end of the century, contingent on fire management practices. Beyond this threshold, irreversible losses and damages are anticipated. Recent research by wildfire experts suggests that current Mediterranean fire management policies, focused on suppression, are destined to fail as they are unfit to account for the non-linear impacts of climate and landscape changes (Moreira et al., 2020).

The primary risks of climate change to forest ecosystems include increased fire danger, shifts in precipitation patterns, altered species compositions, disruptions in carbon sequestration, impacts on water

---

N. Usta Baykal
Natural Sciences Institute, Division of Ecology, Hacettepe University, Ankara, Turkey
e-mail: nurbahar.usta@hacettepe.edu.tr

İ. Bekar
Technical University of Münich, Munich, Germany
e-mail: ismail.bekar@tum.de

resources, and the introduction and spread of invasive species, further threatening forest health and biodiversity. Moreover, the effects of these changes on the adaptive capacity of species and ecosystems remain uncertain, amplifying the urgency for comprehensive understanding and action.

Considering that Mediterranean forests in southern Europe play a vital socio-ecological role, providing crucial ecosystem services that shape economies, cultures, and societies, there is a need for a comprehensive re-evaluation and expansion of the EU Adaptation Strategy (2021) to prioritise climate-related risks to health and ecosystem well-being. We argue that this necessitates a holistic planning approach grounded in multispecies justice, capable of building greater capacity to counter these risks.

To develop a holistic perspective, this chapter draws on a critical literature review co-conducted by its authors in fields as diverse as political ecology, ecological economics, conservation biology, and forest ecology. Further enriching this perspective, an expert workshop in March 2024 brought together a diverse group of 12 specialists, including forest engineers, rural sociologists, ecological economists, environmental education experts, spatial planners, and environmental justice organisers. Additionally, we systematically examined EU strategies on adaptation and forestry to ensure that our approach was grounded in both academic research and real-world practices.

In what follows, we offer a value-based approach to climate risk management and a multispecies justice-focused policy planning process for the Euro-Mediterranean forests. As such, we first present the key debates at the intersection of climate policy and forest management. Subsequently, the discussion shifts to wildfires, highlighting these as a critical issue in the context of climate change and emphasising fire management as one of the key challenges associated with climate change. Offering a value-based approach grounded in IPBES's Nature's Contribution to People (NCP) framework, we conclude with a recommendation to mobilise multispecies justice in climate risk management and environmental planning in Euro-Mediterranean forests.

Changing Façades of Forest Policy in Europe

Forests constitute a key subject in various regulations within the EU's environmental policy. The new EU Forest Strategy for 2030 (European

Commission, 2021a, 2021b) aims to enhance the multifunctional forests in EU territory with a focus on building resilience against the high uncertainty posed by the changing climate. Yet, this policy shift does not proceed uncontested. On the one hand, forest owners and the forest products industry claim that the European Commission did not fully acknowledge the socio-economic sustainability aspect (Eustafor, 2021). On the other hand, nature conservation interests, while admitting that the strategy recognises the need to strengthen the protection, restoration, and biodiversity-friendly forest management, simultaneously accuse the Commission of putting short-term economic gains ahead of other considerations (WWF, 2021). Furthermore, the new Forest Strategy also led to a political cleavage among member states, which became evident in a joint letter issued by the ministers from Austria, the Czech Republic, Estonia, Finland, Germany, Hungary, Latvia, Poland, Romania, and Slovakia. The joint letter asserted that the new forest strategy appeared to favour a "technocratic, one-size-fits-all, and top-down approach developed by the Commission" and demonstrated a lack of comprehension regarding the multifunctional role of forests and the forest-based sector (Köstinger, 2021, p. 2).

These criticisms of the new EU Forest Strategy can be addressed by redirecting the policy focus towards ecosystem integrity and socio-ecological well-being, instead of emphasising the primacy of other interests. The claim that this strategy does not acknowledge the differences among member states can also be interpreted as the negligence and lack of attention to the peculiarities of different and diverse forest ecosystems. EU territory covers 14 different forest ecosystems with 76 sub-types (European Environment Agency, 2006) in which Mediterranean woodlands (forests & shrublands) represent 13% of the total area. Due to an overlap of rich biodiversity with major climate change risks (Giorgi, 2006), the EU plays a crucial role in addressing forest-related global goals (FAO, 2018).

Through guidelines of the Commission such as "Closer to Nature Forest Management", "Primary and Old Growth Forests", "Biodiversity-friendly Afforestation", and "Reforestation and Tree Planting", scientific forestry can contribute to achieving more biodiversity-enriched forests instead of focusing on economic gain. Similarly, a coordinated effort to address the complex challenges to the well-being of forest communities, the needs of the forest industry, and the ambition to enhance ecosystem integrity requires sharper attention to wildfires as the prime example of risks in Euro-Mediterranean forests.

Beyond the Firefighting Trap

Mediterranean forests stand as crucial ecosystems of plant diversity on Earth, hosting around 25,000 species, approximately half of which are endemic to the region. Moreover, these forests offer a multitude of essential benefits and services to society, extending well beyond conventional forest products (Rizvi et al., 2015). The spectrum of vegetation types encompasses diverse landscapes, ranging from forests and woodlands to shrublands, and grasslands. The forest management systems in Mediterranean countries also exhibit a great diversity and unique characteristics, varying from one country to another despite similarities in climate and vegetation. The distinctions are particularly evident in factors such as land ownership ranging from predominantly privately owned western wing of Euro-Mediterranean forests to public ownership dominated eastern flank (Pulla et al., 2013) and the objectives set for ecosystem services. Together, these factors position the Mediterranean as a region with high fire activity, rendering it uniquely vulnerable. Changes in historical fire regimes, driven by climate change and anthropogenic influences, are currently observed and expected to continue. Consequently, the implementation of rapid and comprehensive solutions to address this issue is imperative.

In light of projections indicating a future increase in burned areas (Amatulli et al., 2013), an intriguing paradox emerges when contrasted with the existing trend of decreasing burned areas in European countries (Turco et al., 2016). This shift is likely to be influenced by prevailing fire suppression policies. Expenditure on suppression activities has increased and is expected to continue to do so (Mateus & Fernandes, 2014). However, despite intensive suppression efforts, it is recognised that extreme wildfire events are expected to continue, raising concerns about the long-term sustainability of the resources allocated to suppression activities (Moreira et al., 2020). Thus, this expected increase in extreme wildfire events, regardless of the substantial resources allocated to fire suppression, raises a critical question: *Why do national policies continue to focus on and heavily invest in fire suppression even though such policies fail to address the challenges of a climate-changed future in forests?*

The dominant policy approach in the Mediterranean, which emphasises fire suppression to minimise total burned areas, is expected to be ineffective in the long term for several reasons. For instance, successful implementation of fire suppression policies may inadvertently lead to a "firefighting trap" (Collins et al., 2013). This "firefighting trap" happens when a significant portion of the fire management budget and effort is allocated to fire suppression, with insufficient focus on strategies such as fuel removal to prevent fuel buildup at the landscape scale. Fuel buildup, when coupled with fire management policies that neglect the impacts of climate change results in larger and more intense fires in the long term under extreme fire weather conditions. Moreover, traditional suppression methods are proving inadequate to cope with the size and intensity of these large fires (Fernandes et al., 2016). While these reasons are not exhaustive, they highlight the pressing need for major policy change in Europe, particularly as the significance of wildfires grows not only in Mediterranean Europe but also in other European regions, including northern parts of the continent (Stoof et al., 2024). In recent years, experts have called for concrete measures to address the current challenges highlighting the severe ecological and economic consequences of changes in long-standing fire regimes (Kreider et al., 2024; Mauri et al., 2023; MedECC, 2020; Moreira et al., 2020). These calls should be rapidly acknowledged and integrated into the EU adaptation strategy, emphasising the urgency of transformative action and synergies between mitigation and adaptation (Rayner, 2023).

Towards a Value-Based Understanding of Forest–Climate Interaction

According to The Intergovernmental Science-Policy Platform on Biodiversity and Ecosystem Services (IPBES) NCP framework, the nature–people relationship can be framed in a wide range of different views, where on one extreme, there is a clear separation between humans and nature and on the other, humans and non-humans are deeply interconnected with strong bonds (Díaz et al., 2018). These bonds provide a more direct and suitable framework for a discussion around multispecies approaches. Mediterranean forests promote human well-being by delivering multiple material, non-material, and regulating NCPs, such as regulating watershed hydrology, protecting against erosion, filtering water, providing wood and food resources. They also feature cultural, spiritual, and religious importance for local people and communities

(FAO, 2018). As a result of the multiplicity of different NCPs, policies regarding governance, management, and protection must entail an inclusive, multi-layered, and pluralistic process.

However, despite growing recognition of the importance of incorporating diverse socio-ecological values into policy and decision-making frameworks (IPBES, 2022), contemporary approaches to forest management and governance continue to primarily prioritise market values and concentrate on NCPs that directly impact human well-being economically (Isaac et al., 2024). The implicit hierarchy of values, which is dominated by an anthropocentric perspective that sees humans as separate and, more problematically, superior to other species, results in the mal-governance of forest ecosystems as well as maladaptation to climate impacts.

People, especially forest communities, "have complex and multifaceted interactions with forest ecosystems that go well beyond subsistence" (Bozok, 2024, p. 277). This, in turn, creates a strong relationship between humans and forests insofar as when individuals consider the natural world, they often first think about forests. However, while such sentiments are broadly beneficial for establishing societal legitimacy for forest protection, they may lead to an excessive emphasis on fire suppression to protect forests, resulting in alteration of forest dynamics, and rendering them more vulnerable to larger and more intense wildfires in the long term. Equating forests to untouched nature is also detrimental to broader environmental policy and socio-ecological objectives when other natural areas (e.g., steppes, wetlands, prairies) are considered.

CENTRING MULTISPECIES JUSTICE IN CLIMATE RISK MANAGEMENT

To go beyond the conundrum of forest protection-climate change adaptation, we argue that a multispecies justice lens can provide a helpful way out. Multispecies justice seeks to decentre the human by going beyond human exceptionalism through its focus on cross-scalar and mundane interactions between human societies and more-than-human worlds (Tschakert et al., 2021). Thus, it is defined as "politics for composing a common world that considers the needs and livelihoods of a diversity of human and non-human life" (Jones, 2019, p. 485). This calls for relational, cross-scalar, and careful attention to humans and more-than-humans in Mediterranean forest environments to move beyond a utilitarian and unidimensional understanding of the values of these areas.

A major implication of the multispecies lens to climate risk management is to scrutinise property relations as the basis of justice. By foregrounding the "property" question against privatisation and individualisation, multispecies justice helps re-centre relationality, collective ownership, commoning, and care dimensions (Celermajer et al., 2023). Reframing the web of relations between humans and their non-human environment under climate duress is a key step here. For instance, rather than resorting to catastrophic narratives under increased fire danger in Mediterranean forests, a multispecies approach could help reveal the inherent values of fire-adapted ecosystems (including but not limited to forests and scrubs) to their human and non-human inhabitants. This also calls for a culturally sensitive, gender-conscious, grounded, and relational understanding of fire risk management (often a technical response to naturally occurring phenomena) by not solely resorting to end-of-the-pipe fire suppression solutions. In line with calls for a shift from "matters of fact" to "matters of care" in climate adaptation (Tozzi, 2021), we argue that a rendering of vulnerability as a "state of shared fragility" between humans and more-than-human nature can expand the policy horizon.

Conclusions and Recommendations

Euro-Mediterranean forests stand at crossroads. They face a complex intermesh of stressors such as climate risks amplified by demographic pressures, economic disparities, and societal change. Addressing these challenges demands a specific approach that acknowledges the region's complex interplay of ecological, social, and cultural factors. Such an approach necessitates a transformative shift in policy framing, moving beyond one-size-fits-all strategies and embracing localised solutions grounded in a deep understanding and collaborative action.

An important pillar of this new approach must be rethinking climate risk management policies imposed from above. Instead, a co-creation approach beyond tokenistic participation in policymaking, one that features open dialogue and cooperation with diverse stakeholders, including scientists from social and natural sciences, forestry experts, and local communities is imperative. To carefully assess trade-offs, such co-creation processes may embrace frameworks such as NCPs and multispecies justice as their starting points. Based on our SSH-STEM collaboration experience, we conclude that interdisciplinary and multi-stakeholder endeavours are challenging, especially considering the current divide between social and natural sciences requiring additional effort and time to learn from each other. However, this co-creation and scientific translation process is absolutely crucial to a holistic understanding of the problem. Therefore, such scientific and societal collaborations need to be nurtured further.

Forests are highly valuable, diverse natural ecosystems that justify current fire management and restoration methods. Nonetheless, a new set of policies is required to transform people's relational values under climate impacts, particularly through raising public awareness about effective fire management and restoration methods while preserving instrumental, relational, and intrinsic benefits. Drawing on a more balanced and pluralistic NCP approach in forest governance will not only enhance the efficacy and legitimacy of these policies but will also foster the connection between rights-based approaches to conservation and sustainable use of nature for a good life (IPBES, 2022).

Humans actively make and re-make their environments but so do non-human actors. Vulnerability to wildfires, in this sense, is a conflict-ridden, resistance-generating process that inherently is emotional and thus foregrounds relations of care and interconnectedness between human and non-human nature (González-Hidalgo, 2023). Furthermore, the increase in the size and intensity of fires due to inadequate fire management strategies reinforces this conflict-ridden and resistance-generating process (Kreider et al., 2024). As such, the aftermath of wildfires should not only be thought of as a hurried ecological and biophysical reconstruction effort but also as a psycho-social health process involving grief for lost lives and landscapes (ibid.).

In sum, building a resilient future for the Euro-Mediterranean forests requires embracing contextualised knowledge held by forest communities. Integrating their voices, local ecological knowledge, and concerns alongside scientific forestry approaches can lead to more effective climate risk management strategies that are culturally sensitive and context-appropriate. Such knowledge exchange needs to go beyond mere data extraction; it necessitates genuine partnerships and respect for the diverse ways of being and knowing that have sustained these communities for centuries.

References

Amatulli, G., Camia, A., & San-Miguel-Ayanz, J. (2013). Estimating future burned areas under changing climate in the EU-Mediterranean countries. *Science of the Total Environment, 450–451*, 209–222.

Bozok, N. (2024). Women and forests in solidarity: A multispecies companionship case from the Aegean forests of Turkey. *Gender & Society, 38*(2), 276–298.

Celermajer, D., Schlosberg, D., Wadiwel, D., & Winter, C. (2023). A political theory for a multispecies, climate-challenged world: 2050. *Political Theory, 51*(1), 39–53.

Collins, R. D., de Neufville, R., Claro, J., Oliveira, T., & Pacheco, A. P. (2013). Forest fire management to avoid unintended consequences: A case study of Portugal using system dynamics. *Journal of Environmental Management, 130*, 1–9.

Díaz, S., Pascual, U., Stenseke, M., Martín-Lopez, B., Watson, Z. M., et al. (2018). Assessing nature's contributions to people. *Science, 359*, 270–272.

European Commission. (2021a). *New EU Forest Strategy for 2030.* https://eur-lex.europa.eu/legal-content/EN/ALL/?uri=CELEX:52021DC0572

European Commission. (2021b). *Forging a climate-resilient Europe—The new EU Strategy on Adaptation to Climate Change.* https://eur-lex.europa.eu/legal-content/EN/TXT/?uri=COM:2021:82:FIN

European Environment Agency. (2006). *European forest types: Categories and types for sustainable forest management reporting and policy.* Technical report No 9/2006.

Eustafor. (2021). *The New EU Forest Strategy for 2030 Position of European Forest Owners and Managers.* https://eustafor.eu/uploads/20211004_The-New-EU-Forest-Strategyfor-2030-Position-of-European-Forest-Owners-and-Managers.pdf

FAO. (2018). *State of Mediterranean Forests 2018. Food and Agriculture Organization of the United Nations.* FAO and Plan Bleu, Marseille, France.

Fernandes, P. M., Pacheco, A. P., Almeida, R., & Claro, J. (2016). The role of fire-suppression force in limiting the spread of extremely large forest fires in Portugal. *European Journal of Forest Research, 135*(2), 253–262.

González-Hidalgo, M. (2023). Affected by and affecting forest fires in Sweden and Spain: A critical feminist analysis of vulnerability to fire. *Sociologia Ruralis, 63*(3), 729–750.

Giorgi, F. (2006). Climate change hot-spots. *Geophysical Research Letters, 33*(8).

Helseth, E. V., Vedeld, P., Vatn, A., & Gómez-Baggethun, E. (2023). Value asymmetries in Norwegian forest governance: The role of institutions and power dynamics. *Ecological Economics, 214*, 107973.

IPBES. (2022). *Methodological assessment report on the diverse values and valuation of nature of the intergovernmental science-policy platform on biodiversity and ecosystem services* (P. Balvanera, U. Pascual, M. Christie, B. Baptiste, & D. González-Jiménez, Eds.). IPBES Secretariat, Bonn, Germany.

Isaac, R., Hofmann, J., Koegst, J., Schleyer, C., & Martín-López, B. (2024). Governing anthropogenic assets for nature's contributions to people in forests: A policy document analysis. *Environmental Science & Policy, 152*, 103657.

Jones, B. (2019). Bloom/split/dissolve: Jellyfish, HD, and multispecies justice in Anthropocene Seas. *Configurations, 27*(4), 483–499.

Kreider, M. R., Higuera, P. E., Parks, S. A., et al. (2024). Fire suppression makes wildfires more severe and accentuates impacts of climate change and fuel accumulation. *Nature Communications, 15*(1), 2412.

Köstinger, E. (2021, July 2). *Joint Letter of Ministers responsible for Forestry of Austria, Czech Republic,Estonia, Finland, Germany, Hungary, Latvia, Poland, Romania and Slovakia on the EU Forest Strategy post-2020 Vienna*. Federal Minister. https://skog.no/wp-content/uploads/2021/07/Joint-Letter-on-EU-Forest-Strategy-post-2020.pdf

Mateus, P., & Fernandes, P. M. (2014). Forest fires in Portugal: Dynamics, causes and policies. In F. Reboredo (Ed.), *Forest context and policies in Portugal* (pp. 97–115).

Mauri, E., Hernández Paredes, E., Núñez Blanco, I., & García Feced, C. (2023). *Key Recommendations on Wildfire Prevention in the Mediterranean*. European Forest Institute.

MedECC. (2020). *Climate and environmental change in the Mediterranean Basin—Current situation and risks for the future*. Union for the Mediterranean, Plan Bleu.

Moreira, F., Ascoli, D., Safford, H., Adams, M. A., Moreno, J. M., Pereira, J. M. C., et al. (2020). Wildfire management in Mediterranean-type regions: Paradigm change needed. *Environmental Research Letters, 15*(1), 011001.

Pulla, P., Schuck, A., Verkerk, P. J., Lasserre, B., Marchetti, M., & Green, T. (2013). *Mapping the distribution of forest ownership in Europe* (EFI Technical Report 88, 92 p). https://efi.int/publications-bank/mapping-distribution-forest-ownership-europe

Rayner, T. (2023). Adaptation to climate change: EU policy on a Mission towards transformation? *NPJ Climate Action, 2*(1), 36.

Rizvi, A. R., Baig, S., Barrow, E., & Kumar, C. (2015). *Synergies between climate mitigation and adaptation in forest landscape restoration*. IUCN.

San-Miguel-Ayanz, J., & Camia, A. (2010). Forest Fires. In *Mapping the impacts of natural hazards and technological accidents in Europe: An overview of the last decade* (pp. 47–53).

Stoof, C. R., Kok, E., Cardil Forradellas, A., et al. (2024). In temperate Europe, fire is already here: The case of The Netherlands. *Ambio, 53*, 604–623.

Tozzi, A. (2021). Reimagining climate-informed development: From "matters of fact" to "matters of care." *The Geographical Journal, 187*(2), 155–166.

Tschakert, P., Schlosberg, D., Celermajer, D., Rickards, L., Winter, C., et al. (2021). Multispecies justice: Climate-just futures with, for and beyond humans. *Wiley Interdisciplinary Reviews: Climate Change, 12*(2), e699.

Turco, M., Bedia, J., Liberto, F. D., Fiorucci, P., von Hardenberg, J., Koutsias, N., Llasat, M.-C., Xystrakis, F., & Provenzale, A. (2016). Decreasing fires in Mediterranean Europe. *PLoS One, 11*(3), e0150663.

WWF. (2021). *EU Forest Strategy hampered by shortsighted interests*. https://www.wwf.eu/?4040816/EU-Forest-Strategy-hampered-by-shortsighted-interests

Open Access This chapter is licensed under the terms of the Creative Commons Attribution 4.0 International License (http://creativecommons.org/licenses/by/4.0/), which permits use, sharing, adaptation, distribution and reproduction in any medium or format, as long as you give appropriate credit to the original author(s) and the source, provide a link to the Creative Commons license and indicate if changes were made.

The images or other third party material in this chapter are included in the chapter's Creative Commons license, unless indicated otherwise in a credit line to the material. If material is not included in the chapter's Creative Commons license and your intended use is not permitted by statutory regulation or exceeds the permitted use, you will need to obtain permission directly from the copyright holder.

CHAPTER 12

Conclusions

*Ester Galende Sánchez, Alevgul H. Sorman,
Violeta Cabello, Sara Heidenreich,
and Christian A. Klöckner*

Abstract This book presents ten interdisciplinary contributions addressing key policies of the European Green Deal. The chapters emphasise the need for inclusive participation of all actors, integrating justice in policy design and implementation as well as tackling controversial issues such as deep sea mining and carbon dioxide removal. Overall, the book advocates for 1) Reimagining knowledge transfer, and emphasising mutual learning between the global North and South; 2) Strengthening the integration of diverse knowledge systems to develop robust, people-centric, transformative climate policies; 3) Seeking climate

E. Galende Sánchez (✉) · A. H. Sorman · V. Cabello
Basque Centre for Climate Change (BC3), Leioa, Vizcaya, Spain
e-mail: ester.galende@bc3research.org

A. H. Sorman
e-mail: alevgul.sorman@bc3research.org

V. Cabello
e-mail: violeta.cabello@bc3research.org

S. Heidenreich
Department of Interdisciplinary Studies of Culture, Norwegian University of Science and Technology, Trondheim, Norway

© The Author(s) 2024
E. Galende Sánchez et al. (eds.), *Strengthening European Climate Policy*, https://doi.org/10.1007/978-3-031-72055-0_12

justice and global equity into climate policy to avoid negative impacts beyond EU borders; 4) Promoting accountability and transparency in all decision-making processes, and; 5) Embracing justice and diversity - of voices, of contexts, of knowledges, and of disciplines to tackle one of the most complex collective action problems to date, the climate crisis which involves us all.

Keywords Transformation · Climate Policy · Inclusivity · Justice · Reimaging

In the last few years, we have witnessed significant advancements in climate action. All major emitters (e.g., the EU, the USA, and China) have established strategies and plans for decarbonisation, and the EU is set to become carbon neutral by 2050 at the latest. However, the latest Emissions Gap Report reveals that climate policies are still insufficient to keep our planet and all its living beings in a safe and thriving space (UNEP, 2023). Therefore, we need to embark on a **continuous journey of reimagining**: questioning long-held beliefs and paradigms, and increasing the ambition and transformative potential of these policies. The book you have in your hands presents a contribution to this aim from ten interdisciplinary and diverse teams, addressing key policies of the European Green Deal, such as the EU Adaptation Strategy, the Circular Economy Action Plan, the EU Forests Strategy, the Energy Efficiency on Buildings Directive, as well as more controversial issues such as deep-sea mining or carbon dioxide removal.

While addressing different themes, there is an underlying agreement across the chapters that the solutions to the climate and ecological crises require the **inclusive and meaningful participation** of all actors. "Leaving no one behind" is one of the key pillars of the EU Green

---

e-mail: sara.heidenreich@ntnu.no

C. A. Klöckner
Department of Psychology, Norwegian University of Science and Technology, Trondheim, Norway
e-mail: christian.klockner@ntnu.no

Deal and there is ample consensus on its importance (see for instance, Galende-Sánchez & Sorman, 2021; Ostrom, 2014; Perlaviciute & Squintani, 2020). However, there is still a long way to go before this slogan becomes a reality on the ground. Quoting Shery Arnstein (1969, p. 216): *"The idea of citizen participation is a little like eating spinach: no one is against it in principle because it is good for you"*. Concretely, García Mira et al. (Chapter 8) argue that, despite the implementation of policies such as the Just Transition Mechanism, communities from coal regions do not feel sufficiently engaged in the decarbonisation processes, as there is often a lack of methods or knowledge for how to implement truly inclusive decarbonisation processes. Authors show the need to empower local and disadvantaged communities in the design of compensatory measures, as well as in regional visions, plans, and narratives, and how this can be achieved through the combination of SSH with STEM methodologies like system models or simulations to even out knowledge inequalities if they are implemented smartly. Likewise, Abdel-Fattah et al. (Chapter 9) underline the need for equitable participation in Maritime Spatial Planning that recognises the importance of all relevant stakeholders and pays attention to power dynamics, ensuring a just distribution of positive and negative social and environmental impacts. Plaček et al. (Chapter 10) bring to life a collaborative workshop setting to assess ethical dilemmas and trade-offs in the transition to a low-carbon construction sector and emphasise the importance of recognising the diversity of stakeholders and of public sector leadership.

Another aspect needed to strengthen EU climate policy is further incorporating **justice** in the design and implementation of policies and strategies, including distributive, recognitional, restorative, procedural, and epistemic justice. To support these efforts, Bobadilla et al. (Chapter 5) argue for the need to further incorporate local knowledge into the EU Adaptation Strategy and present an indicator to measure its epistemic justice. In the international context, the analysis conducted by Portugal-Pereira et al. (Chapter 2) unveils the risks of EU mitigation policies relying on carbon dioxide removal projects in the Global South and how a Eurocentric focus might fulfil the EU's climate commitments at the expense of vulnerable populations, perpetuating colonial patterns. Instead, Portugal-Pereira et al. argue for EU climate policy to further focus on reducing domestic emissions. In a similar vein, Suarez-Visbal et al. (Chapter 3) highlight the need to make vulnerable and marginalised voices part of the conversations about the governance of the transition towards circularity in the textile sector and avoid and repair the harm to workers in the Global South who make most of the clothes consumed in the EU. Both Portugal-Pereira et al. and Suarez-Visbal et al. (Chapters 2 and 3) show that EU policy cannot be seen within a European context only, as it inevitably has global consequences that need to be considered.

Nevertheless, achieving justice also requires taking into account the diversity of the people and ecosystems within the EU. Lara-García et al. and Turhan et al. (Chapters 6 and 11) highlight specific vulnerabilities of Southern Europe (i.e., heatwaves and wildfires), underscoring that "one size does not fit all" in EU policies (for an in-depth discussion on this topic regarding EU research policies please see Varjú et al., 2023). Dealing with these complexities essentially requires both the sensitivity of SSH methodologies and the power of STEM methodologies to understand, describe, and simulate complex socio-technological systems. In line with cross-sector complexities, Lara-García et al. (Chapter 6) advocate for the recognition of the climate crisis as a public health problem and highlight the links between inequalities, exposure to heatwaves and health risks. This chapter calls for the EU Renovation Wave to consider an integrated approach in prioritising funding through interdisciplinary assessments, where again STEM and SSH need to go hand in hand to build infrastructure that supports living in a hotter world. Turhan et al. (Chapter 11) discuss the EU Forest Strategy for 2030 and the challenges it faces to adequately deal with increased climate-related risks across a diversity of forest ecosystems, especially in the Mediterranean. As extreme wildfire events are expected to continue, traditional fire suppression management strategies need to be revised to avoid the "firefighting trap" (Collins et al., 2013). The authors call for a value-based approach that acknowledges forest–climate relations and activates synergies between mitigation and adaptation strategies within the EU Adaptation strategy.

The solutions to the climate crisis require us to reimagine and transform our cultures, our societies, our economies, our technologies, our scientific approaches, and our political systems. Therefore, climate policies need to **question long-held beliefs and practices**. In this sense, Menatti et al. (Chapter 4) propose to rethink the current (neocolonial) trends of knowledge transfer from Global North to Global South and instead focus on **mutual learning** and translation of knowledge and experiences across different geographies. Concretely, the authors present different examples of how Southern Europe could benefit from translating long-term practices in India to face heatwaves and extreme temperatures. The authors suggest this shift requires a different conceptualisation of adaptation as *"processes which proactively engage with disrupting climate events and involve adopting situated and relational long-term practices relating people and their ecological, social and historical environments"*. Seeland et al. (Chapter 7) advocate the need for societal indicators in carbon accounting to promote social sustainability and end the prioritisation of economic indicators such as economic growth (i.e., GDP Gross Domestic Product). Many authors have similarly argued this notion for years and the EU Parliament hosted a Conference on the topic (Parrique, 2022; Raworth, 2018). Turhan et al. (Chapter 11) propose a multispecies rather than an anthropocentric approach to deal with climate risks, thus recognising the innate value of both humans and non-human living beings and placing us as part of nature, not apart from it (Fremaux & Barry, 2019; Kopnina & Washington, 2020). Lastly, Suarez-Visbal et al. (Chapter 3) highlight the need to challenge current business models and practices sustained in overproduction and overconsumption and the need to go to the root causes of the socio-ecological crises.

Overall, this book has combined insights from an extensive set of disciplines, including Political Science, Biology, Architecture, Geography, Philosophy, Marine Sciences, Engineering, Psychology, Energy Systems Modelling, and much more. Despite evidence that scientific knowledge and STEM are not enough to deal with social-ecological crises, there is still insufficient integration of diverse knowledges and disciplines into policy development (Crowther et al., 2023; Foulds & Robison, 2018; Turnhout, 2024). Therefore, this book aims to underline the value of SSH for climate policy and governance, and how interdisciplinary efforts of STEM and SSH jointly unlock results in innovative and robust methodologies and policy recommendations.

For instance, Menatti et al. (Chapter 4) advance an interdisciplinary framework for knowledge integration based on conceptual comparative analysis and creative games to uncover epistemic differences and develop collaborative skills for mutual learning, translation, and, ultimately, coproduction. Lara-García et al. (Chapter 6) show how geography and architecture can be combined to develop multi-criteria assessments of vulnerability to heat in urban environments, integrating social, biophysical, and building indicators. García Mira et al. (Chapter 8) combine SSH and STEM leading to different, innovative outcomes that neither of the perspectives would have been able to achieve on its own: When a lignite power and heating plant in a region in Greece was to be shut down, the initial plan was replacing the heating need by gas boilers which was also embraced by the citizens. However, in a combination of SSH citizen engagement and energy systems modelling and simulations, alternative fossil-free scenarios were developed. In sum, interdisciplinary approaches together with the inclusion of different knowledge systems bring a more holistic outlook to complex climate challenges.

Climate policy is one of our most powerful tools in the most crucial years to address the climate crisis and ensure the well-being of present and future generations. For climate policies to be genuinely transformative, they need to embrace justice and diversity—of voices, of contexts, of knowledges, and of disciplines. We are facing one of the most complex collective action problems to date, and **the solutions necessarily involve all of us.**

References

Arnstein, S. (1969). A ladder of citizen participation. *Journal of the American Planning. Association, 35*(4), 216–224.

Collins, R. D., de Neufville, R., Claro, J., Oliveira, T., & Pacheco, A. P. (2013). Forest fire management to avoid unintended consequences: A case study of Portugal using system dynamics. *Journal of Environmental Management, 130*, 1–9.

Crowther, A., Foulds, C., & Robison, R. (2023). *A review of the Climate-Energy-Mobility landscape through 10 Social Sciences and Humanities literature briefs.* SSH Centre.

Foulds, C., & Robison, R. (Eds.). (2018). *Advancing energy policy: Lessons on the integration of Social Sciences and Humanities.* Springer International Publishing.

Galende-Sánchez, E., & Sorman, A. H. (2021). From consultation toward co-production in science and policy: A critical systematic review of participatory climate and energy initiatives. *Energy Research & Social Science, 73*, 101907.

Kopnina, H., & Washington, H. (Eds.). (2020). *Conservation: Integrating Social and Ecological Justice.* Springer International Publishing.

Ostrom, E. (2014). Do institutions for collective action evolve? *Journal of Bioeconomics, 16*, 3–30.

Parrique, T. (2022). *Ralentir ou périr: L'économie de la décroissance.* Éditions du Seuil.

Perlaviciute, G., & Squintani, L. (2020). Public Participation in Climate Policy Making: Toward Reconciling Public Preferences and Legal Frameworks. *One Earth, 2*(4), 341–348.

Raworth, K. (2018). *Doughnut economics: Seven ways to think like a 21st-century economist*. Random House Business Books.

Turnhout, E. (2024). A better knowledge is possible: Transforming environmental science for justice and pluralism. *Environmental Science & Policy, 155*, 103729.

UNEP. (2023). *Emissions Gap Report 2023: Broken Record—Temperatures hit new highs, yet world fails to cut emissions (again)*. United Nations Environment Programme.

Varjú, V., Tagai, G., Cabello, V., Sorman, A. H., Robison, R., Foulds, C., Bálint, D., Galende Sánchez, E., Zindulková, K., Pálné Kovács, I., et al. (2023). *Supporting the Social Sciences & Humanities across Southern and Central & Eastern Europe: A position statement for international climate, energy and mobility research*. SSH CENTRE. https://sshcentre.eu/wp-content/uploads/2023/11/Position-estatement_English.pdf

Open Access This chapter is licensed under the terms of the Creative Commons Attribution 4.0 International License (http://creativecommons.org/licenses/by/4.0/), which permits use, sharing, adaptation, distribution and reproduction in any medium or format, as long as you give appropriate credit to the original author(s) and the source, provide a link to the Creative Commons license and indicate if changes were made.

The images or other third party material in this chapter are included in the chapter's Creative Commons license, unless indicated otherwise in a credit line to the material. If material is not included in the chapter's Creative Commons license and your intended use is not permitted by statutory regulation or exceeds the permitted use, you will need to obtain permission directly from the copyright holder.

Afterword 1. Where Do We Go from Here? SSH Inquiries into Crucial HOW Questions in Climate Change Policies

Irmelin Gram-Hanssen *works as a Senior Researcher at the Western Norway Research Institute in Sogndal, Norway, where she focuses on climate change adaptation as an entry point into sustainability transformations. She is committed to generating actionable knowledge that simultaneously helps solve problems and supports the potential for thriving people and the planet.*

The entangled social-ecological crisis of our time, such as climate change, biodiversity loss, and social inequality, are complex and require multiple perspectives and knowledge sources to be understood and addressed. To this end, interdisciplinarity is indispensable. While the STEM fields and their associated methods are skilled at inquiring into the WHAT of climate change, SSH are crucial when it comes to inquiring into HOW. The lack of progress within climate action indicates that while we should continue asking questions of WHAT, the questions of HOW are increasingly pressing. HOW do we—as societies, governments, social movements, businesses, and individuals—make the changes necessary to halt dangerous climate change? While technology is part of the answer, it is highly insufficient. Technologies of various kinds are nothing but tools. It is how we pick them up and use them, to what aim, with what logic, for whose benefit, that matters for the results we will generate. This is where the collaboration between SSH and STEM fields is crucial, and the present book is full of excellent examples of how interdisciplinary research engages such complexity.

In the context of interdisciplinary climate change research, the SSH are particularly important in pointing to the complexities involved when identifying challenges and possibilities—that the perspectives we take impact our ability to respond. This is exemplified where Seeland et al. zero in on the logics underpinning carbon-reducing interventions, calling for a shift from the predominant economic-based rationale to a socio-benefit rationale focused on equity. Similarly, Menatti et al. call for a reframing of adaptation from a mere coping with adverse impacts of climate conditions to a situated and relational long-term process involving people and their ecological, social, and historical environments.

The SSH are also key in highlighting issues of justice and equity within climate change—noting winners and losers, identifying root causes of vulnerability and asking whose voices count. Examples of this include Suarez-Visbal et al., who highlight the need for democratic inclusion of diverse voices within the creation of a circular textile strategy and the prioritisation of alleviation mechanisms as opposed to corrective measures to enable textile actors in the Global South to transform their practices sustainably. Similarly, Bobadilla et al. argue for the integration of local knowledge into EU policymaking as well as the advancement of epistemic justice in dealing with climate adaptation.

Finally, the SSH are instrumental in connecting global change research and policymaking to the local contexts where specific knowledges and wisdom will need to be leveraged to generate transformative change. For instance, García-Mira et al., write about the importance of equipping key actors with the transformative capacities to support the development and implementation of regional visions, plans, and narratives at the local level. Similarly, Turhan et al. suggest that the livelihoods of forest communities, their local ecological knowledge, and traditional practices should be considered in tandem with scientific forestry practices.

In sum, SSH perspectives have been crucial in understanding climate change not as a technical problem, but as an adaptive challenge that is highly political and cultural. Yet, where do we go from here?

Both science and experience tell us that we need to take immediate action to mitigate and adapt to climate change, even though we do not know the full picture. Yet, it is crucial that such action not only avoids doing harm but enhances sustainability across social-ecological systems, geographies, and cultures. We must hurry, but we must hurry slowly to avoid perpetuating existing inequalities in the process.

The most crucial role for SSH within interdisciplinary research going forward may not necessarily be to identify all the ways that our actions are insufficient or have unintended consequences. Such critique is important, but critique in and of itself has limited value. Rather, going forward we will need to practise our ability to keep multiple realities and modalities within view at the same time. More specifically, such modalities include simultaneously solving problems while realising the potential for a better present and future for people and planet. It includes honouring the specificity of challenges and solutions in any given place while simultaneously identifying the ways in which our efforts can include everyone everywhere. And it includes working out in the world while simultaneously working within ourselves to be able to notice our blind spots, question our assumptions, and to get out of our own way so that we not only speak and teach about transformation but become an integral part of that transformation ourselves. This takes knowledge and skills, but maybe more importantly, it takes humility, courage, and compassion. Because interdisciplinary is itself a practice of multiplicities, it is a powerful place to practise such abilities.

Afterword 2. Cross-Disciplinary Thinking to Rise to the Challenges of Global Systemic Risks

Ajay Gambhir *is Director, Systemic Risk Assessment, at the Accelerator for Systemic Risk Assessment (ASRA), and a Visiting Senior Research Fellow at the Grantham Institute, Climate Change and the Environment, at Imperial College London. His current research focuses on how to understand and assess systemic risks both within and between systems including climate, environment, economy, food, health, energy, technology, finance, and others, across scales and geographies, considering both human societies and ecosystems. He has almost two decades of experience working on energy and climate change, including roles in the UK government and academia.*

The world is facing planetary-scale hazards—we have far exceeded our thresholds of safety on six of nine quantified measures of biophysical and biochemical processes that regulate Earth's life-support systems for humanity and many other species (Richardson et al., 2023), including around climate change, land and freshwater system changes, ocean acidification, and chemical cycles. Although expressed in separate metrics, these processes are interrelated and stem from our Great Acceleration (Steffen et al., 2015) in material production and consumption. Moreover, this physical analysis does not show our social and economic pathologies, which stem from the same roots (Vineis & Gambhir, 2023). In summary, the scale is planetary, which means transboundary in terms of countries and geographies, and the scope is inter-systemic, implicating economic, societal, environmental, and technological systems. This calls for truly global-scale, cross-disciplinary, systemic thinking.

It is therefore encouraging to see the array of analyses presented in this volume that apply such thinking in both their scope and methods. Portugal-Pereira et al.'s assessment of the environmental and social implications of the EU sourcing its bioenergy for carbon dioxide removal from Brazil demonstrates how damaging such an action could be. The policy recommendations, at a time when carbon offsets and markets loom large in international discussions of equity, finance, and green colonialism, are clear: do more emissions reductions domestically, and source less from abroad. Seeland et al. argue that even within the EU, a purely economic, market-based focus on emissions permit trading ignores social considerations, and that a much broader consideration of factors is required to ensure effective and just emissions reductions.

In fact, justice is a recurring theme throughout these analyses. Suarez-Visbal et al.'s multidimensional justice lens, drawing from circular economy and post-growth, reveals how the EU's circular economy strategy around textiles has multiple inadequacies. These include a lack of recognition of the diversity of lived realities of textiles workers across the world, as well as the fundamental forces of fast fashion and overproduction that drive unsustainability. As asserted by García-Mira et al., truly just pathways are unlikely, unless the top-down policymaking approach that still dominates in the EU is replaced with a much more collaborative approach, including citizens and community groups. Furthermore, as argued by Turhan et al., justice is not purely about consideration of different societal groups, but also incorporates the welfare of non-human species, as elucidated when considering forest protection in the continent.

Justice and equity in stakeholder participation are increasingly highlighted as central to achieving an ethical transition to a sustainable future. Plaček et al. use an interdisciplinary approach to assessing the complex ethics of transitioning to low-carbon construction in the EU, encompassing knowledge from ethics, policy, sociology, engineering, and architecture advocating balanced stakeholder engagement, collaboration, transparency, and accountability. Abdel-Fattah et al. also highlight the importance of equitable participation in maritime spatial planning—a process that needs to be improved so as to redress power and knowledge imbalances across stakeholder groups.

Nevertheless, regardless of how effective and just the transition to a low-carbon future may be, one of the painful realities of our climate emergency is that many climate impacts are now inevitable, and adaptation is essential—but here again, participation and justice are central.

For example, Bobadilla et al. highlight how a drive towards "epistemic justice" (the inclusion of comprehensive knowledge from all relevant stakeholders), is of paramount importance in the implementation of effective and just adaptation policies, which may need to be enacted at different speeds and scales in different localities and contexts throughout the EU. Such enactment also needs better data, as demonstrated by Lara-García et al.'s multivariate analysis of factors affecting heat vulnerability in European residential buildings highlighting the need for more comprehensive data to help focus building renovation efforts towards greater adaptive capacity. As well as enhanced data, experience from other regions could be vital. For example, Menatti et al.'s analysis of how Indian communities have managed heatwaves in the past, through building design, use of tree cover in urban spaces, and empowerment of women to train households in the use of energy efficiency and solar reflective methods, advocates for a careful translation of these lessons to European contexts.

The sheer range of issues to consider when assessing how to drive towards a safer future for humanity and other species is daunting. Disciplinary silos, geographically narrow and bounded analyses, and nation-state interests placed far above the needs of the planetary commons all still predominate. But the chapters in this volume show encouraging signs that researchers are rising to the challenges of our multidimensional, interconnected world, in an era of global systemic risks. Such systemic research and analysis—driven by transdisciplinarity to bridge SSH with STEM—are powerful tools to reveal unintended harms and hazards and the appropriate responses to transform towards genuine sustainability and justice.

References

Richardson, K., Steffen, W., Lucht, W., Bendtsen, J., Cornell, S. E., Donges, J. F., ... & Rockström, J. (2023). Earth beyond six of nine planetary boundaries. *Science Advances*, 9(37), eadh2458.

Steffen, W., Broadgate, W., Deutsch, L., Gaffney, O., & Ludwig, C. (2015). The trajectory of the Anthropocene: The great acceleration. *The Anthropocene Review*, 2(1), 81–98.

Vineis, P., & Gambhir, A. (2023). Social inequalities and the environmental crisis: Need for an intergenerational alliance. *Frontiers in Public Health*, 11, 1226961.

Afterword 3. Justice as the Foundation of European Climate Policies: A Future that Serves All of Us

Rebecca Thissen works for Climate Action Network (CAN) International as Global Lead for Multilateral Processes. CAN is the widest climate network globally, bringing together more than 1900 members across the globe dedicated to pushing for an ambitious climate justice agenda. She is a lawyer specialising in International Public Law and Human Rights. Over the course of her career, Rebecca has been focusing on the questions of equity and justice in climate and energy governance, specifically strengthening the connection between human rights, social justice, and climate change.

To combat global warming, we urgently need to put in place cross-cutting and systemic solutions that will enable us to tackle the root of the problem, rather than treating the symptoms with short-term band-aids. Developing sensible, fair, and equitable solutions will require forcing our thinking out of its traditional shackles and taking it to a more holistic, transversal level. This is why my initial interest in this book was the nature of the project itself: interdisciplinary, multi-actor, cross-sectoral, and focusing on social innovations. We cannot do without anyone, and silos are our first enemies.

Secondly, I want to echo the importance of first-party participation in the implementation of climate policies. Social dialogues and engagement mechanisms are regularly discussed, particularly in the context of just transition policies, as illustrated by García-Mira et al. However, participation rights (also enshrined in the Paris Agreement) are threatened, reduced, or even annihilated in many contexts. Moreover, this needs to

© The Editor(s) (if applicable) and The Author(s) 2024
E. Galende Sánchez et al. (eds.), *Strengthening European Climate Policy*, https://doi.org/10.1007/978-3-031-72055-0

be situated in a broader context where the fundamental human right to protest is being increasingly criminalised and environmental defenders are facing mounting risks and repression. All in all, it is important to realise that the current state of civic space is shrinking fast, all over the world. Hence, participation policies must not only be intentional, but also proactive in their approach, equipping key actors with the right capacities and ensuring genuine inclusion of the various stakeholders, starting with the most vulnerable, as shown by Lara-García et al., Turhan et al., and Plaček et al.

My third observation concerns the international dimension of certain chapters and the link to European climate and energy policies. The inclination not to consider or to minimise their impact beyond the European borders, particularly in the Global South, is a major concern. As highlighted by Portugal-Pereira et al. and Suarez-Visbal et al., understanding these global interconnections and how domestically-oriented policies can have consequences elsewhere is key. This is a warning for the future of climate governance and international relations in general: no economy, and certainly no developed country, can afford not to adopt a global and solidarity-based vision of its climate and energy policies. As a matter of fact, Europe urgently needs to rebuild equitable relations with its partners in developing countries, rather than disguise its so-called "strategic partnerships" behind green neo-colonialism logics. I also particularly liked Menatti et al. approach suggesting that the EU should look to the Global South for inspiration on adaptation, given their track record and resilience in the face of extreme weather events. This echoes a lesson of humility. In the current context, the EU cannot present itself as a champion of climate action, trumpeting the merits of its climate goals and policies, while at the same time failing to live up to its international commitments and remaining deaf and obstructing the solutions emanating from the Global South, in all areas of climate action.

In the end, I believe that the underlying question we should be asking ourselves while reading this book is whether justice can be placed as the foundation of current and future European climate policies. If the answer is yes, it needs to be understood that climate and energy policies cannot only be about GHG or carbon removals. The climate crisis is rooted in a much more complex and deeper system, and therefore requires complex and deep alternatives. To avoid a future that will only serve a very few and will violate our collective right to a healthy environment, we need to look for alternatives that promote human rights, international solidarity, and effective and informed participation as compasses.

Afterword 4. Interdisciplinary Perspectives to Reimagine Systems for a Sustainable and Just Future

Beth Doherty is a youth climate advocate and final-year law student at the University of Cambridge. She recently attended COP28, and is working to draft legal agreements protecting the Arctic. She has also co-led a special thematic report to the UNCRC on the link between children's rights and the climate crisis.

Anna Pellizzone is a natural scientist lent to journalism and social sciences. Her professional life has always been at the intersection of different disciplines. She is a freelance professional in citizen and stakeholder engagement, experienced in designing and running participatory processes in Research and Innovation.

The climate crisis cannot be solved by one sector alone, and justice must be achieved through systemic, cross-society change. This book therefore explores how interdisciplinary research can support policy to reimagine our systems and build a brighter, more sustainable and just future. In response to this challenge, a key thread tying the chapters together is climate justice. This invites consideration of what justice means, how its meaning and application can vary in different contexts, and what can be done practically to achieve a truly just transition. One crucial aspect of climate justice is recognising and engaging with those communities which have been historically excluded from decision-making. These communities are often those most vulnerable to and affected by the climate crisis, as explored by García Mira et al. Given the climate crisis' roots in colonial and extractive practices, which persist today, climate justice also calls on us to decolonise our approaches to climate action, discussed in the context of the EU's carbon removal strategy in Brazil by Portugal-Pereira et al.

In recognising the flaws in our current systems, we must also have the bravery to reimagine what our world can look like. An element of this explored throughout this book is a shift in how we measure progress and assign value. As suggested by Seeland et al., we must shift our focus away from the current economic-based rationale and instead towards societal sustainability. Economic growth means little if it delivers great wealth for the few, at the expense of the lives and livelihoods of many, especially those who are most vulnerable to the effects of climate change. By shifting our focus, we change how we assign value, and choose to change our framework from one which prioritises pure economic growth to one which prioritises sustainability and a regenerative relationship with the natural world. This also requires us to change from a purely anthropocentric view of the world, and instead recognise our role as part of nature as stated by Turhan et al.

While research and innovation across STEM and SSH disciplines have a role to play in addressing the climate crisis, they must engage societal actors (e.g., citizens; policymakers at different scales) in that process or they won't succeed. This is particularly important due to the complexity and interconnected aspects of the climate crisis, which call for a mosaic of solutions. The chapters tackle this head-on and provide inspiration on how knowledge—including tacit knowledge[1]—distributed in society can be valued, integrated, and translated into actions to reach common goals (i.e., mitigation and adaptation to climate changes), with insights on how to deal with three issues: (1) how to crowdsource and coalesce knowledge from different disciplines, actors, and places; (2) how to maximise this knowledge and turn it into policies and solutions for the social benefit; (3) how to fairly monitor and evaluate the impact of climate-related actions. This aligns not only with mission-like approaches, but also with EU knowledge valorisation policies, in which "knowledge valorisation" is defined as *"the process of creating social and economic value from knowledge by linking different areas and sectors and by transforming data, know-how and research results into sustainable products, services, solutions and knowledge-based policies that benefit society"* (OJEU, 2022, 3). In this process, including all societal actors is imperative.

The evidence stemming from the chapters is that the most powerful tool we have to integrate knowledge, needs, and opportunities towards a common direction is to bring people together around the table. Whether to co-design policies, to co-create solutions, to build multidimensional indicators, we need sound dialogue and participation. Experiments on this front are flourishing,[2] and while a lot more still needs to be done, this book is a valuable contribution and source of inspiration in that process.

NOTES

1. Tacit knowledge is "any knowledge that cannot be codified and transmitted as information through documentation, academic papers, lectures, conferences or other communication channels" (OJEU, 2022, 3).
2. The EU H2020 MOSAIC project is a valuable example of knowledge valorisation through co-creation within the EU Mission Climate-Neutral and Smart Cities.

REFERENCE

OJEU. (2022). *Council Recommendation (EU) 2022/2415 of 2 December 2022 on the guiding principles for knowledge valorisation*. Council of the European Union. https://eur-lex.europa.eu/legal-content/EN/TXT/PDF/?uri=CELEX:32022H2415eu. Accessed 9 May 2024.

Index

A
Adaptation, 2, 4, 6, 36–40, 42–45, 50–52, 57, 58, 63, 68, 71, 127, 130–132, 138–140, 145, 146, 151, 154, 158
Agriculture, 4, 10, 16, 83, 104

C
Circular economy (CE), 4, 22, 138, 150
Climate change, vi, ix–xi, 5, 6, 11, 17, 23, 38, 44, 50, 56, 57, 62, 64, 71, 76, 78, 100, 127, 128, 131, 145, 146, 158
Collaboration, vi, vii, 2, 10, 45, 58, 78, 90, 113, 115, 116, 120, 122, 132, 145, 150
Communities, x, 5, 10, 15, 17, 25, 38, 41, 42, 44, 52–56, 65, 80, 81, 89, 90, 92–94, 105, 126, 128, 131, 133, 139, 150, 151, 157
Consumption, 11, 24, 25, 69, 71, 92, 95, 149

D
Decision-making, 5, 6, 28, 30, 54, 68, 78, 83, 101, 103, 107, 120, 131, 157
Development, v, vi, 5, 15, 17, 23, 38, 51, 59, 91, 95, 101, 105, 113, 141, 146

E
Energy, v, vi, x, 5, 11, 16, 38, 51, 52, 57, 62–64, 66, 69, 71, 77, 78, 80, 89, 91–93, 95, 100, 104, 112, 142, 151, 154
Ethics, 3, 53, 113, 117, 119, 150
EU Green Deal, v, 2–4, 139

G
Global South, 3, 4, 10, 11, 16, 17, 26–30, 36, 37, 44, 45, 139, 140, 154
Governance, v, xi, 3, 6, 14, 23, 51, 95, 106, 131, 133, 139, 141, 154

Greenhouse gas emissions, 2, 9, 11, 13, 17, 77
Greenhouse Gas (GHG), 13, 77, 154

H
Health, vi, 5, 36, 37, 41, 44, 62–66, 70, 71, 77, 83, 127, 140
Heatwaves, x, 5, 36, 37, 40, 42–44, 63–65, 68, 71, 140, 151

I
Inclusivity, 6, 81
Innovation, vii, 17, 39, 42, 84, 95, 116, 117, 153, 158
Interdisciplinary, vi, x, 3, 4, 7, 17, 24, 26, 36–38, 42, 44, 45, 52, 58, 61, 71, 83, 84, 101, 107, 113, 132, 138, 141, 142, 146, 147, 153, 157

J
Justice, 3–5, 11, 23–25, 29, 30, 37, 42, 44, 45, 49–53, 55, 57–59, 78, 102, 127, 131, 132, 139, 146, 150, 157

L
Local knowledge, x, 4, 51–53, 55–57, 59, 139, 146
Low-carbon, v–vii, 82, 111–115, 119–122, 139, 150

M
Mitigation, xi, 14, 16, 30, 63, 82, 130, 140, 158
Mobility, v, vi, 81

N
Net-zero, 3, 77, 83

P
Participation, 6, 30, 53, 55, 91, 95, 96, 100, 106, 132, 139, 150, 153, 155, 159
Policy, vi, vii, x, 2, 4, 6, 16, 23, 27–29, 37, 39, 43, 45, 51, 52, 58, 59, 78, 80, 82, 84, 93–95, 100, 106, 115, 119, 121, 127–129, 131, 132, 140–142, 157
Public engagement, vi

R
Renewable energy, 88, 91, 93, 95, 100, 112
Resilience, 26, 41, 51, 69, 154
Resource use, 2

S
Social justice, 37
Stakeholders, 3, 6, 26, 30, 51, 55, 56, 58, 89–96, 100, 101, 105–107, 112–118, 120, 121, 132, 150, 154, 157
Sustainability, 5, 23–25, 39, 42, 78, 82, 84, 112, 113, 119, 121, 122, 128, 129, 145, 158
Sustainable development, 100

T
Technology, 11, 13, 17, 77, 113, 122, 140, 145
Textile, 3, 4, 23–25, 27–30, 140, 146, 150
Transdisciplinary, 3, 52
Transition, v–vii, x, 2, 5, 11, 24, 25, 28, 78, 82, 83, 88–93, 95, 96, 105, 107, 112–116, 119, 121, 139, 150, 153, 157

V
Vulnerability, 5, 27, 50, 56, 62–65, 67, 94, 132, 133, 141, 146, 151

W
Well-being, ix, 5, 22, 27, 81, 83, 128, 130, 142

SPRINGER NATURE

GPSR Compliance

The European Union's (EU) General Product Safety Regulation (GPSR) is a set of rules that requires consumer products to be safe and our obligations to ensure this.

If you have any concerns about our products, you can contact us on ProductSafety@springernature.com

In case Publisher is established outside the EU, the EU authorized representative is:

Springer Nature Customer Service Center GmbH
Europaplatz 3
69115 Heidelberg, Germany

The manufacturer's authorised representative in the EU is Springer Nature Customer Service Centre GmbH, Europaplatz 3, 69115 Heidelberg, Germany. If you have any concerns regarding our products, please contact ProductSafety@springernature.com

Printed and bound by CPI Group (UK) Ltd, Croydon, CR0 4YY

23/03/2026

02076447-0002